FUTURE ROBOTS SOCIAL POSSIBLE DEVELOPMENT

JOHN LOK

Made with ♥ on the Notion Press Platform
www.notionpress.com

Copyright

Contents

Preface

Nowadays, many businessmen or marketing research professional hope to apply different methods to predict consumer behaviors in order to know what will be future market activities and market changes to help them to choose to implement what kinds of marketing strategies more accurately. The methods include economic environmental change prediction method, consumer individual psychological change prediction method, micro or macro behavioral economic environmental change prediction method, marketing environmental change prediction method etc. different kinds of methods which can be applied to predict how consumer behavioral changes to influence whose behavioral consumption to the manufacturer .

This book aims to explain why and how future artificial intelligent technology (big data gathering method) can be applied to assit businesses to predict why and when and how consumer behavior changes. I shall explain why traditional psychological and statistic and marketing methods are applied to predict consumer behaviors, human's judgement and analytical effort will be worse to compare AI machine's judgement and analytical effort. Also, I shall indicate different business organizations why they apply AI big data gathering method to help them to design any questionnaires (surveys) questions which will be more valid and useful to conclude human's questionnaires (surveys) design questions method.

This book has these two research questions need to be answered: Can apply (AI) learning machine predict consumer behaviors? Can (AI) learning machine replace human marketing research method, e.g. survey or human psychological and micro and macro economic methods to predict consumer behaviors more accurate?

In my this book first part, I concentrate on indicate whether any artificial intelligence (AI) tools will be one kind of good consumer behavioral prediction method to be choose to apply to predict consumer behaviors. I shall indicate some examples, cases to give reasonable evidences to analyze whether (AI) tools will be one kind suitable tool to be applied to predict when and how consumer behavioral changes. If (AI) can be one kind tool to attempt to be applied to predict when and how consumer behavioral changes. Will it replace other kinds of methods to predict consumer behaviors? Does it have weaknesses to be applied to predict consumer behaviors, instead of strengths? Can it be applied to predict consumer behaviors depending on any situations of only some situation? Finally, I believe that any readers can find answers to answer above these questions in this book.

In my this book second part, I shall explain why and how human can possible apply (AI) tool to predict consumer individual emotion. I shall indicate case studies to explain how consumer individual better or worse emotion how to influence whose consumption behavior in different situation. Finally, I shall indicate evidences to conclude how and why (AI) tool that can be used to predict consumer individual emotion and it will have direct relationship to influence consumption behavior, as well as how (AI) tool can assist businessmen to judge whether what reasons case the customer does not choose to buy its product, it is possible because the product high price factor, poor product quality or poor staff service performance or attitude etc. different factors to influence the consumer decides to choose to buy the other product consequently, when the (AI) tool can confirm consumer has good or bad emotion to judge what factors are the causes his decision making at the moment.

Readers can understand why and how (AI) tool can be attempt to be applied to predict customer emotion and it can influence positive or negative consumption behavior to the product clearly in this part.

In my this part, I concentrate on indicate whether any artificial intelligence (AI) tools will be one kind of good consumer behavioral prediction method to be choose to apply to predict consumer behaviors. I shall indicate some examples, cases to give reasonable evidences to analyze whether (AI) tools will be one kind suitable tool to be applied to predict when and how consumer behavioral changes. If (AI) can be one kind tool to attempt to be applied to predict when and how consumer behavioral changes. Will it replace other kinds of methods to predict consumer behaviors? Does it have weaknesses to be applied to predict consumer behaviors, instead of strengths? Can it be applied to predict consumer behaviors depending on any situations of only some situation? Finally, I believe that any readers can find answers to answer above these questions in this book.

In my analysis, I conclude marketing development or marketing change trend will be influenced by consumer behavioral change model or attitude factor. Finally, I hope my readers can give opinions to make judgement to evaluate my opinions whether is right or wrong in this research topic.In part four, I shall indicate sample case study to judge whether it is possible to apply (AI) tool to attempt to help businesses to predict consumer behaviors in different business environment.

Prologue

Table of contents

5.3 Can (AI) big data gather data to predict when
climate will change to influence poor travelling
behaviors?

How can apply (AI) provide travelling businesses with better-informed decisions ? p.154-174

6.1 Future travel consumption behavior

6.2 What is push and pull factors to influence any traveler who chooses where is whose preferable travelling destination ?

6.3 Why can expectation, motivation and attitude factor influence travelling behavior?

What is (AI) deep learning techniques to
forecast travelling environment
behavioral consumption ? p.175-190

7.1 Environmental travel consumption prediction

7.2 What methods can predict future travel behavioral consumption ?

7.3 How to apply advanced traveler information systems (ATIS) to predict future travelling behavior?

7.4 How can online tourism sale channel influence traveling consumption of behavior?

7.5 How can analyze activity based travel demand ?

7.6 What is actively based patterns of urban population of travel behavioral prediction method?

7.7 What is trip based versus activity based approaches?

7.8 Can apply (AI) big data gathering method predict senior age will be main travelling target?

7.9 IS (AI) big data gathering method a better psychological method to compare human marketing research method predict travel behavioral consumption?

7.10 How can apply (AI) digital channel
(big data gathering method) predict
travelling consumer behaviors?

Reference

What factors can influence travel behavioural consumption p.191-220

Prediction travel behavioral consumption from psychology view and computer statistic view.

Whether climate change can influence travelling behaviours.

8.1 Market method predicts future
travel consumption behavior p.241-243

Whether individual habitual behaviour can influence travelling behaviour: e.g. renting
travel transportation tools

How to determine future travel behavior from past travel experience and perceptions of risk and safety for the benefits to travel consumers?

What is push and pull factors to influence any
traveler who chooses where is whose preferable travelling destination.

Why expectation, motivation and attitude factor can influence travelling behaviour.

8.2 What methods can predict future travel behavioural consumption

How to use qualitative of travel behavioural method to predict future travel consumption.

How to apply advanced traveler information systems (ATIS) to predict future travelling behaviour.

How does online tourism sale channel can influence traveling consumption of behaviour.

Actively based patterns of urban population of travel behavioural prediction method.

What is trip based versus activity based approaches?

Why senior age will be main travelling target.

Psychological method to predict
travel behavioural consumption.

Reference

How (AI) predicts MTR passenger
behavior p.221-235

How does MTR (Mass Train Railway) need to
consider route design location of choice by (AI)
marketing research survey method?

Why MTR underground train transportation
needs to know passenger behavior
Why route choice can influence passenger
behavioral choice

Why trip time reliability and crowding factors
can influence MTR passenger choice.

Main barriers influence artificial intelligence
consumer behavioral prediction

10.1(AI) digital data gather technology predicts food consumer behavior's main barriers

10.2The challenges of (AI) big data gather shaping
the future of retail for consumer industries

How does artificial intelligence
influence future working changing
in automation employment and
productivity aspects?
Is artificial intelligence possible
to replace labor ?
Can (AI) technology replace human
labour nature of work?
Why can artificial intelligence satisfy
human needs?
Is artificial intelligence one good choice
for human future technological benefit?

What is relationship between
(AI) and economy growth?
How can artificial intelligence technology influence economy?
Can (AI) influence gloabl economy
growth?
How can artificial intelligence
impact global economy growth?
How can (AI) influence GDP of high income
countries in the next ten years?
How can artificial intelligence
impact on workplace?
What is the relationship
between (AI) and (CRM)?
Can (AI) technology
impact on customer relationship
management (CRM) ?
What is relationship between
(AI) and digital job nature causes?

How can (AI) technology influence digital
economy? P.256-270

What is the relationship between
(AI) and global digital economy development?

Could work activities in China be
automated making in the nation with the world's largest automation potential?

How does (AI) technology influence
the future of employment change?

How can artificial intelligence impact
global economy growth?

Why will (AI) technology grow economic
development ?

How can (AI) technology impact to global
economic and social and psychological
changes?

Will (AI) technology influence digital
economy change to manufacturing
industry ?

What is artificial intelligence potential
benefits and ethical considerations?

How can (AI) technology influence to
global health care economy
development?

Reference

First part Factors influence householder energy using behavior at home

Chapter 3

AI influences householder
electricity energy consumption behavior

Environmental impacts of householder
greenhouse gas electricity energy p.271-290

The effect of house space occupancy
and building characteristics on
householder electricity energy use

How to help low income household
earners to reduce not essential
electricity energy expenditure
spending at homes
Factors influence householder energy
efficient consumption behaviors
at homes p.291-310
How artificial intelligence impacts energy
consumers using behaviours

How to apply (AI) technology to improve
energy efficiency and better climate change
and the security of energy supply as well
a resource efficiency?

The relationship between consumer
psychology and behavior

ONE

(AI) PREDICTION CONSUMER BEHAVIOR TOOL

How can artificial intelligent tools predict consumer behavior in vehicle market

What is (AI) consumer behavioral prediction tool? How any why will (AI) tool assist manufactures to attempt to predict consumer behavior before and after consumption occurrence? First, I shall indicate how to apply (AI) tool to predict vehicle product consumer behavior case example.

Nowadays, many vehicle manufacturers hope their vehicles can attract to vehicle buyers to choose to buy their vehicles. However, there are many different brands of vehicles to provide to them to choose, so the vehicle market competition is very serious.

How to judge their different kinds of vehicle price which is reasonable acceptance to attract vehicle buyers to choose to buy the brand of vehicle manufacturers' any kinds of vehicles, e.g. fast speed sport style vehicles, comfortable and slow speed common cars, for four passengers common small size or more than four passengers common large car size? How to evaluate the vehicle prices issue is important factor to influence vehicle buyers' choices. Either if the brand of vehicle price is too high to compare brands, it will influence many vehicle buyers choose to buy other brands' vehicles or if the brand of vehicle price is too low, it will influence vehicle buyers feel this brand's vehicle's quality is worse to compare to other vehicle brands' similar vehicle products.

Thus, if the brand of vehicle manufacturers can predict how to design vehicles which can attract many vehicle buyers to choose to buy whose any vehicle products. What are future vehicle buyers' favorable vehicle styles? Then, the vehicle manufacturer can concentrate on manufacturing the kind style of vehicle products to sell already. It will reduce its vehicle manufacturing investment risk.

How to apply (AI) tools to predict vehicle buyers' behavioral consumption model? Whether artificial intelligent tools can predict automotive buyers' behavioral consumption model and predict future trend. In fact, automotive brands and dealerships are facing an increasingly competition when attempting to manually gathering the vast quantities of data required to create customer focused programs that increase retention, ultimately new sales and service automotive business. Building a based on that client's intrinsic needs and interests to any kinds of automotive vehicles at any given time. This is especially true in the automotive industry where the time span between purchases is measured in years. Because vehicle buyers would not like often to change their old vehicle to another new one. So, their decisions to buying another new vehicle, the time is usually after one year, even longer time. Hence, it seems any vehicles won't be frequent consumption products to the owned at least one vehicle family consumers (vehicle buyers). Hence, how to predict vehicle consumers' taste or preferable which styles of vehicle choices issues is very important. If the vehicle manufacturers can not manufacture any attractive vehicles to sell easily in this year. Then, it will lose time, money in this year because it won't know when the owned least one vehicle users or non-owned any vehicle users who will decide to buy one new vehicle or change another new vehicle ensure. The different brand vehicle dealers will possible wait more than one year to attract them to buy their vehicles if their styles are not attractive to compare other brands of vehicle competitors.

However, artificial intelligence and machine learning can help any vehicle manufacturers to find solution to solve patterns in highly to solve patterns in highly complex data-sets that are beyond the capability of a human brain, and then building and automatically acting on the customer insights it generates.

Given the automotive customer need for individualized communications, this technology is positioned to become a critical component of any successful vehicle retailer's domestic or/and overseas vehicle markets. How can vehicle manufacturers and retailers use (AI) to enhance their vehicle marketing campaigns? How will (AI) affect their vehicle sale marketing strategy? What criteria would they use when selecting on (AI) solution?

Vehicle consumers today are able to quickly access different brands of vehicle information, research vehicle products and reviews, negotiate prices and compare one vehicle brand or retailer to another resulting of the brands of vehicle customers. At the same time, the rise of " big -data mining", wearable devices that track user's every move and preference and greater contextualization in advertising and social media has resulted in consumer expectations of individualized. Thus, it seems that (AI) tools can be used to gather " big-data" and then they can make human's mind to analyze how to design kinds of vehicles to satisfy vehicle buyers' needs.

As automotive vehicle marketers can apply (AI) tools to achieve messaging strategies to meet the needs of this new generation of informed vehicle consumers, using data from a variety of sources to move from a variety of sources to move from mass- messaging to more personalized messages aimed at particular vehicle buyer segments, e.g. fast speed sport vehicle buyer segment, slow speed comfortable small size or large size of buyer segment. However, when 90% of vehicle marketers believe having a single vehicle buyer view is important, only 6% have achieved it.

However, one of the main issues vehicle marketers facing is the lack of capacity to efficiently sift through and analyze the massive vehicle buyer amounts of data required to create vehicle buyer individualized vehicle customer experiences easily. This is especially difficult for automotive dealers, the long periods between purchase cycles, and the highly considered nature of the vehicle purchase means that each vehicle dealer needs to not only track a large number of potential vehicle customers for an extremely long period of time, but each of those vehicle customers will generate a huge amount of different kinds of vehicle behavioral consumption data as they research their next vehicle purchase. However, by choosing the right (AI) technological tools and programs , vehicle dealers can solve this big data gathering challenge into a major advantage.

For Forrester vehicle brand example, vehicle consumers have more power over the Forrester vehicle brand's reputation than ever before. Mayne, L. (2014) indicated that Forrester calls this new (AI) tools is the " age of the vehicle customer", a 20 year business cycle in which the most successful vehicle enterprises will reinvent themselves to systematically understand and serve increasingly powerful vehicle consumers. To win in this new age, Forrester declares companies must become vehicle customer obsessed and the only sustainable competitive advantage is knowledge and engagement with customers, such as (AI) gathering data knowledge.

Thus, the biggest challenge vehicle businesses currently face is not the collection of a large quantity of vehicle consumer data, but what to do with that data once they have it. Even at a large vehicle data research firm, the data sets are often too big for a single analyze, or even a team of analysts to sort through and draw conclusion from. However, enter artificial intelligence and machine learning , an efficient technology solution that can continuously find patterns in highly complex data sets that are way beyond the capacity of a human brain and then automatic drive action based on the customer insights is generated.

What is (AI) machine learning tool? Machine learning is a type of (AI) that learns from data and is not explicitly program. Think Amazon, face book. Machine learning serves up relevant content based on an individual vehicle purchase behavior and experiences. More simply, machine learning is a computer program that can learn relationships between data, subject those learnings to errors functions, and then learn from its errors. The program in effect, trains itself.

Lee, T. (2016) explained that "Thus, (AI) tools can learn deep a more advanced branch of machine learning inspired by how our brain's nervous function, has also been found to be especial effective in identifying patterns from data."

When this way sound is complicated from a vehicle dealer perspective, the implementation of a marketing program driven by artificial intelligence can take care of these tasks in an automatic vehicle fashion with little to no manual intervention required from the staff at time vehicle stores.

In practice at a vehicle dealership, the program will continue track vehicle customer behavior online, merging that data with any offline source (like CRM or DMS data) and then analyze this aggregated vehicle buyer data set to predict what vehicle customer may be shopping for and what information they might like to relevance from different kinds style of vehicle design photos.

1.1 Why can (AI) be applied to predict consumer behaviors?

Artificial intelligence refers to complex in vehicle market, machine learning that posses the same characteristics of human intelligence and that have all our sense, all our reason and think just like human do. Besides, machine learning is the practice of using algorithms to collect and examine data, learn from it, and then make a determination or prediction about something in the world.

The machine is " trained" using large amounts of data and algorithms that give it the ability to learn how to automatically perform a task with increasing accuracy. Otherwise, deep learning is primarily based on artificial neural networks inspired by our understanding of the biology of human's brains.

Deep learning breaks down tasks in ways that enables machines to assist us with increasingly complex tasks, driverless cars, better preventive healthcare and more accurate product recommendation (including vehicle recommendations). So, such as why (AI) technology can be applied to predict how vehicle consumer behavior changes to bring to judge whether vehicle consumer will like what kinds of vehicle styles next year. Then, vehicle manufacturers can gather overall vehicle consumer data to analyze and conclude the more accurate vehicle design direction for next year any new design vehicle manufacturing products.

Thus, (AI) machine learning can help vehicle manufacturers to solve how to design any new vehicle products challenge. A vehicle is both one of the most important and carefully considered purchases the majority of people will ever make in their lifetime. It is also a purchase that tends to be fundamentally tied to a person's identify and view of themselves. As the same time, vehicle consumers changing lifestyles result in changing vehicle needs, e.g. the young sport car enthusiast matures into the family driver.

Automotive dealers need to remember that vehicle customers and prospects are individual human beings with risk, complex and ever-changing lives factors, these factors will influence every vehicle consumer why who feels has vehicle purchase need, and how who choose to buy the first vehicle if who decided to buy the first vehicle.

The (AI) technological customer behavioral prediction tool seems to be the best vehicle salespeople in the world are those that know every one of their vehicle customers. Their likes and dislikes which style of vehicle design, preferences and changing tastes to vehicle choices. The capacity of the human brain, however, limits us from achieving this type of vehicle sales and frequent turnover at vehicle dealerships often results in the further loss of vehicle salespeople along with their vehicle customer relationships and knowledge. In this competitive vehicle environment, machine learning enables platforms to assist the vehicle sales team by tracking the vehicle consumer behaviors of each vehicle customer, learning and memorizing their preferences and predicting their future vehicle purchase needs.

Finally, I recommend that for a vehicle dealerships marketing platform to make their customer engagement efficient and fully-functional, I should be able to: applying (AI) tools to track every vehicle customer behavior across the web, connecting to a society of data sources, CRM, DMS, third-party, web vehicle brands, social email, click etc., aggregating and accurately cross-reference data from a variety of sources, leveraging this data to drive insights on a mass scale, as well as on an individualized basis, driving actions and automatically direct customer engagement via multiple channels based on where each customer is in their individual lifecycle.

1.2 How can (AI) provide businesses with better-informed decisions

I shall explain how (AI) technology can provide businesses with better-informed decisions to drive top-line growth, deliver meaningful experience for customers and smooth their path along the consumer journey. The widely understood definition of (AI) involves the ability of machines or computers to learn human thinking, reasoning and decision-making abilities.

A Narrative science study in 2015 year identified that (AI) was being used primarily in voice recognition, machine learning virtual assistants and decision support. This study also highlighted the many branches of (AI) and that techniques and their definition are used interchangeably. It is possible that (AI) can be used to gather big data , then to analyze to help businesses to predict consumer behaviors. For example, one of the most common techniques is machine learning, where algorithms are used to perform tasks by learning from historical data. Another growth branch of (AI) is natural language procession.

However, during 2017 year, search engines will begin to factor additional behavioral data into prediction of customer behavioral results, such as the user's history of searches and locations and previously captures conservations. Artificial intelligence will use this information to power predictive search results, e.g. predictive future consumer's choice behavioral processing for any kinds of businesses.

Predictive search will improve the quality of search results, and provide new insights into consumers' behavior and the moments which matter to them. Search will give recommendation into tailored how consumer individual choice in consumption process. Several of the largest online platforms already use machine learning to improve predictive consumer behavioral search results.

For example, Google's rank brain technology adds research by understanding the context in which the consumer has entered it. Over time, rank brain will learn further from user behaviors Amazon's DSSTNE (pronouned destiny) learns from shoppers' purchasing habits and consumption behavior to offer better product recommend actions, which Amazon can offer before a consumer has entered anything into the search bar. However, this technology is not independent of human input. For example, Google engineers will periodically retain the rank brain system to improve the models it uses. For another example, in 2016 year , Apple computer revamped its photos app to allow consumers to search for specific items in the phots, they want to find, not just dates and locations. Each photo that an intelligent phone or intelligent pad user takes goes through 11 billion computations, so that photos can understand exactly what is the photography.

It seems that in future, (AI) machine learning will allow search to evolve even further. Search engineers will deliver refined recommendations to their business users and use less human input to predict consumers' needs. For IBM computer example, it indicated 90% of the data that exists today has been created in the last two years. This huge explosion of data gives brands the opportunity to quickly spot and react to the latest trends, fashion and fads among its clients and potential clients. This will allow companies to better engage with younger consumers, who gain influence access to the latest trends, and use the brands. They associate with to help define who they are as individuals. Thus, brands have to identify and make use of them before consumers move on, but the vast quantity of data available makes. This a resource-intensive task. For next example, Lesara, a based online clothes store, uses this machine learning to inform its product decision often gathering information from internal and external sources. When its trends -spotting shoes. Lesara has a range of over 20 styles and sells hundreds of pairs a day. It focus on giving consumers, the very latest trends allow Lesara to develop on average of 50,000 new items each year. It compared to 11,000 old items each year. Thus, (AI) brain seems to human brain to own analytical ability to predict consumer behaviors.

For another example, Lesara is one online clothes store, uses machine learning decisions after gathering information from internal and external sources. One of its most popular products, shoes with LED started life when its trend spotting software flagged up a blogger wearing similar shoes. Now Lesara has a range of over 20 styles and sells hundreds of pairs a day. Its focus on giving consumers the very latest trends allows Lesara to develop an average of 50,000 new items each year, compared to 11,000 for its competitor Lara. it seems (AI) machine learning can help Lesara business to predict what kinds of shoes design or style that shoe consumers will prefer choose to buy in future shoe market trend. Thus, Lesara can predict shoe consumers' taste successfully and it can manufacture many attractive style of shoes. (AI) machine learning can gather global past shoe consumer's shoe shopping experiences, then analyzes to make conclusion to give lesara recommendation successfully. This will make the experience more enjoyable for shoe consumers and allow Lesara to advert whose different new style or design of shoes to deliver them move relevant messages by understanding the context of the experience.

However, (AI) machine learning will have this risk who manufacturers need to concern if they applied this technology to predict consumer behavior. It is on sample consumers' privacy issue, in order to avoid complaint chance occurrence. However, machine learning can tie this data together to identify which f the billions of devices are being used by individual consumers. This helps brands understand how consumer engagement and actions can be attributed to different messages in different contexts and at different time. So, machine learning can help brands to build confidence to promote their products by any advertisement channels. When, this new (AI) machine learning technology can conclude how to design their products to be the most attractive, due to it has more accurate to predict consumer behaviors to compare human themselves prediction judgement effort. It seems that (AI) machine judgement effort is more accurate to compare to human judgment effort.

For example, google is moving away from cookies and using logged in data to track and make to users. It plans to expand the scope of the brand lift tool from online video. Thus, consumers are responded will to shippable context, finding it persuasive and easy to navigate by (AI) machine learning decision. For example, fashion brands can aggregate their You tub videos and blogs into a mobile context marketing experience, such as brand centric context into a personal shopping activity gives the shopper an experience, who are likely to remember and tell their friends about any new style of products design promotion from these internet advertisement channels after (AI) machine learning tools' styles of product design recommendation.

What is (AI) deep learning techniques to forecast environment behavioral consumption

The (AI) deep-learning technology leads to performance enhancement and generalization of artificial intelligent technology. It influences the global leader in the field of information technology has declared its intention to utilize the deep-learning technology to solve environmental problems, such as climate change. So, it will help agriculture farming businesses can raise any plant food: vegetable, fruit, rice which grow up very easily if farmers can apply

(AI) deep-learning technology to solve environment problems to influence their plant food grow. If the whole year seasonal change is very good and it is suitable for any plant food to grow in farming land easily, e.g. rain is enough and soil is enough for any plant food to grow in the farm lands. Then, fruit, rice, vegetable etc. agriculture businesses will have much beneficial attribution to global farmers.

The question is how to use deep-learning technologies in the environmental field to predict the status of pro-environmental consumption. We predicted the pro-environmental consumption index based on Google search query data, using a recurrent neural network (RNN model). To certify the accuracy of the index, we compared the prediction accuracy of the RNN model with that of the ordinary least square and artificial necessary network models. For example, the RNN model predicts the pro-environmental consumption index better than any other model. we expect the RNN model to perform still better in a big data environment because the deep-learning technologies would be increasingly as the volume of data grows. So, deep-learning technologies could be useful in environmental forecasting to prevent damage caused by climate change to influence any rice, vegetable, tomato, potato, fruit etc. different plant food grow in any countries' farming land easily.

For South Korea example, over 800 government agencies spent 2.2 trillion Korea won on eco-products in 2014 year. However, green products are rarely purchased outside these agencies. This phenomenon occurs because there is a gap between consumer attitudes and behavior , that is environmental attitude is a major factor in decision making vis-a-vis the consumption of " green" food and services (Jorea Ministry of Environment, 2015). Therefore, it is necessary to understand those consumer attitude, that will lead to sustainability-conductive behavior and consumption.

2.1 Environmental consumption prediction

Recently, many researchers have studied pro-environmental consumption and household indexes as well as suicide rate predictions using messages posted by internet users on Google trend, Tweets etc. channel. Whether can environmental consumption be predicted by (AI) deep-learning technological internet channel? How can impact the pro-environmental consumption attitudes of green policies? Korea scientists estimated pro-environmental attitudes using search query data provided by Google trend and confirmed through regression analysis, that pro-environmental attitude has a positive correlation with the pro-environmental attitude index. They also explained that environment-friendly attitude of residents plan an important role in policy making. In the past, most household consumption indexed were calculated through surveys, but (AI) deep-learning technological tool " big data" have recently gained research attention (Lee et al. 2016).

It seems that (AI) deep-learning technology can help agricultural export countries' farmers , e.g. US, UK, Canada, New Zealand, Australia, Japan, China, India etc. they can predict environmental behavioral consumption to any rice, tomato, potato , fruit, vegetable etc. plant food consumers. The beneficial advantages to them include as below:

(a) Assuming they know their countries' weather, when it has less rain to cause drought or when it has more rain in any seasonal time in the year. They can choose not to grow any kinds of above these plant food to avoid loss.

(b) They can make any kinds of above these plant food price raising after their prediction of these bad seasonal time to cause their plant food shortage supply challenge. Because these plant food consumers' demand number is more, but the supply of these above plant food supply number is less. However, due to they had predicted when the bad seasonal time can not allow them to grow these above plant food before. So, they have enough time to grow many these above plant food number in predictive good seasonal time to prepare to supply to their plant food import countries' plant food consumers to eat. Thus, these predictive environmental consumption plant food export countries can raise their plant food price to sell to them. When, the other non-pre-predictive environmental consumption plant food export countries can not supply any one of those plant food to them to eat, due to the bad climate to cause them can't grow any one of these plant food to export to sell.

Thus, (AI) deep-learning technology can be applied to predict how to raise the plant food supply number in order to raise price to the import plant food countries consumers to eat, due to they feel difficult to buy these plant food to eat in the bad climate seasonal time in whole year.

(c) (AI) deep-learning technology can help climate scientists to find what reasons cause their countries; rain sudden increases or cause their countries' rain sudden decreases. After its gathering data analysis, it can assist climate scientists to find solution methods to attempt to control the rain level can be right falling down level to let agricultural export farmers who can grow their plant food to sell to agricultural import countries in whole year.

(d) The agricultural export countries' farmers can apply (AI) deep-learning technology to help them to choose whether growing which kinds of plant food in that whether climate time to earn more plant food consumption number more easily.

Due to the agricultural countries climate will often change, for example, tomato, potato, rice, fruit etc. plant food can be adapt to grow in more rain time, but vegetable can not be adapt to grow in more rain time. If farmers can apply this technology to predict when it will have move rain or when it will have less rain to fall down in their countries. Then, they can choose to grow which kinds of plant food number more, in the suitable seasonal climate time in order to raise plant food growing number productivities to supply to sell to satisfy any agricultural food import countries' demand effectively.

(e) (AI) deep-learning technology can help agricultural import countries to solve agricultural food shortage challenge in long term. When this technology can be popular to base applied by the agricultural plant food export countries. It will solve global agricultural food shortage challenge. For example, when one agricultural export countries' farmers can popular accept to apply this technology to predict when to grow which kinds of plant food more to rise number productivities to sell. e.g. vegetable, fruit, rice Besides another agricultural export countries' farmers can also accept to apply this technology to predict when to grow plant food, e.g. potato, tomato to raise number productivities to sell. Then, they can concentrate on growing the specific kinds of plant food in order to raise the specific plant food number productivities in every seasonal change time every month. Then, global agricultural plant food supply must be raised, due to these predictive environmental change farmers can know who ought grow which kinds of plant food to sell to raise number productivities.

2.2 How can apply (AI) digital channel to predict consumer behaviors?

(AI) digital channel can be applied to help businesses to evaluate whether how much the product price is the most attractive to persuade consumers feel it is the most reasonable price to sell. It helps consumers to feel which brands of products which ought change the price to let consumers to choose to buy the brand of product. It can be applied to predict whether how many consumer numbers can be increased or decreased when the brand of product's price is variable. It aims to give opinions to help any brand of product manufacturers or sellers to judge whether which price is the most reasonable to let consumers to accept to choose to buy the brand of product in popular.

Thus, (AI) price measurement technology can be preference to be applied online communication ecommerce and mobile phone internet platform aspect. As businesses can enter their past products prices data and past customer number data into computer or mobile. Then, (AI) price measurement technology can gather these data to analyze these product prices and past customer number to compare their prices variable changing range level to find their price variable difference to measure to make conclusion about every product's price variable changing will influence how many customer number increase or decrease changing to choose to sell their different kinds of products more accurate. Then, (AI) price measurement software will help them to analyze all past price variable changing data to compare whether which price range can let customers to feel it is more reasonable and attractive to influence them to choose to buy the product among different brands of product choice.

Because any product's price is one important factor to influence consumers to choose to buy the product, instead of quality, durability, shape, appearance, color, brand familiarity etc. factors. Any online businesses with a focus on Asia should considerate (AI) customer care, and virtual shopping experience, whereas is Europe and North America still value face-to-face and/or real human interaction over (AI) or virtual worlds.

For example, Amazon publish has applied (AI) price measurement technology to help authors to decide how much every different topic of e-book or paper book price, it can attract the largest number of readers to buy. Any one author only needs to type whose book name to Amazon publish author himself/herself Amazon website. Amazon publish (AI) price measurement learning machine will help them to auto-calculate and judge how much e-book or paper book price is the most attractive and the most reasonable in order to increase reader number to buy their e-books or paper books to read. So, (AI) online price measurement machine will gather past similar book names and past every similar book readers' reading times and the number of readers to give opinions to let every author to judge whether his/her very new e-book or paper book ought charge how much price to the e-book or paper book which can attract many readers to choose to buy. Although, it is not ensure that the e-book or paper book price must let readers to feel it is the most reasonable price to choose to buy in reader's view point. However, it has other factors to influence readers' choice to buy the e-book or paper book, e.g. whether the book content is attractive to public, the author's familiarity, the book's page is enough or not to satisfy readers to read etc. factors. But, instead of all these extra factors to influence readers to choose to buy the book to read. (AI) price measurement learning machine can real give opinions to every author to let them to judge the e-book or paper book different price range whether is too high to influence readers to choose to buy to read or tool low to influence readers feel it is possible poor content book to compare other similar content books. Thus, (AI) price measurement machine can help authors to predict every reader's reading behaviors or reading experience and reading habit from online channel in short time easily. The author only enter the book name to let Amazon publish price measurement machine to check, it will follow past reader's reading habit and reading experience to judge whether the similar all book topic sale record to judge how much price is the reasonable price to attract many readers to buy the book.

Hence, (AI) can be applied to digital channel to help businesses to predict consumer behavior in the future. In the future, mobile/smartphone, laptop, desktop will be most frequent used ecommerce channels to develop online business. So, (AI) can be also applied to these platforms to gather data to make analysis to help businesses to predict consumer purchase behaviors popularly. Due to , ecommerce is popular to global, so digital online and instore channels can be one good channel to let (AI) learning machine to make platform to gather past every online consumer purchase (buying) experience data to help businesses to build brand personality and having a responsible, positive impact on society.

To apply (AI) learning machine technology to understand customer online purchase behavior, it will raise business e-commerce successful chance: For example, (AI) learning machine can help businesses to gather data to analyze to determine whether short-term or long-term signals in the online consumer behavior that indicate higher purchase intents to let every online business to know. (AI) learning machine can find that online users with long-term purchasing intent tend to save and click through on more content.

However, as online users approach the time of purchase their activity becomes more topically focused and actions shift from saves to searches from online consumption channel. Then, (AI) learning machine will further find that the brand product purchase signals in online behavior can exist weakness before an online purchase is made and can also be traced across different online purchase categories. Finally, (AI) learning machine synthesize these insights in predictive models of online user purchasing intent to the brand of product. Taken together, it's work identifies a set of general principles and signals that can be used to model online user purchasing intent across many online content discovery applications. Thus, (AI) learning machine can help online businesses to gather any online users' click online behaviors data to judge whether there are how many online users will choose to find their online business websites to make final decisions to buy their products from online channels. Then, it will give opinions to help the online businesses to let it to judge whether what are the important website factors will help its online business to attract many online consumers, e.g. designing unattractive website issue, online unattractive product photos issue, unclear website color issue, unclear website advertisement message, contents and words impressions issue, lacking image movement frequent attractive seeing issue etc. different website factors. Thus, online digital channel will be one good choice to apply (AI) learning machine to help businesses to predict consumer behaviors.

2.3 Can apply artificial intelligent learning machine " big data" gathering method to predict manufacturers' behavioral performance ?

In consumer view point, can they apply (AI) learning machine to predict manufacturers' behavioral performance to judge whether whose products are value to buy. Nowadays, (AI) and big data are reshaping the risk in consumer privacy. For example, consumers want to hide their willingness to pay just as firms want to hide their real marginal cost, and buyers have less favorable information, say a low credit shore, prefer to withhold it just as sellers want to conceal poor product quality. So, it implies that it is possible (AI) learning machine can help customers to gather any manufacturers‘ past sale performance, e.g. how many complaints or appreciation from clients, product quality etc. sale data to let consumers to make judgement whether it is value to buy to compare other competitors. So, it has risk to the poor product quality of manufacturers. Otherwise, it has benefits to the good product quality of manufacturers. It also implies all manufacturers' privacy is not protected or secret when (AI) learning machine is popular to be used to predict manufacturers‘ behaviors by consumers.

Information economists suggest that both buyers and sells have an incentive to hide or reveal private information, and these incentives are crucial for market efficiency. Data technology that reveals consumers type could facilitate a better match between product and consumer type, and data technology that helps buyers to assess product quality could encourage high quality production.

Thus, (AI) big data technology can also assist consumers to gather different manufacturers' data to compare what their advantages and disadvantages of their products are. Then, consumers can make comparison to choose which brand of product is the suitable to whom to buy in these more choice consumption market. (AI) learning machine will gather similar brand their products‘ data to analyze to make conclusion to let consumers know or feel to make final judge to find what advantages or disadvantages of these sample brands of similar products' comparison from internet. On the other hand, it means that manufacturers can gather consumers‘ past purchase behaviors or purchase experience from (AI) big data gathering method to record and analyze to give opinions to let manufacturers to know what reasons or factors influence consumers choose not to buy their products from internet.

(AI) big data gathering consumer behavior prediction method can give these benefits to manufacturers and consumers both, such as: New concerns arise because (AI) technological advance which have enables reducing cost of collecting, storing, processing and using data in mass quantities extend information beyond a single transaction. These advances are often summarized by the big data, it means charge volume of transaction-level data that could identify individual consumers by itself or in combination with the datasets.

The popular (AI) takes big data as in input in order to understand, predict and influence consumer behavior. Modern (AI) is used by legitimate companies, could improve management efficiency motivate innovations and better match demand and supply. But (AI) in the wrong hand, also allows the mass production of fraud and deception. Since , data can be stored, traded and used long after the transaction. Future data use is likely to grow with data processing technology, such as (AI) big data gathering consumer and manufacturer behavioral prediction method from internet channel.

Thus, future (AI) big data learning machine can also help consumers to choose the best brand of manufacturer's products among different brands of manufacturers products choice to compare their past sale performance from internet. They can apply (AI) big data statistic method to gather all different manufacturers' similar products past sale data to compare their advantages and disadvantages to make the best decision to choose to buy which brand of product is the most suitable to them to buy to use. It seems (AI) big data can also help consumers to predict any manufacturers' manufacturing behaviors or manufacturing performance whether they are improving their product quality or are deteriorating their product quality. Thus, (AI) big data tool is also important to help customers to predict future the different brands of manufacturer performance will have improvement in possible.

Thus, I believe that artificial intelligent "big data" gathering method can be suggested to be applied to attempt to predict consumer behavioral changes in global business environment, the reasons are as below:

On the consumer's beneficial hand, Consumers can apply this method to attempt to gather any global manufacturers data to be analyzed by this artificial intelligent learning system. Then, it analyzed all the different brands of specific similar product manufacturer' data to compare what are the range of the best past manufacturing history and sale data to the group of best manufacturers, and what are the range of the better past manufacturing history and sale data, and what are the range of the good past manufacturing history and sale data, and what are the range of the common past manufacturing history and sale data. Finally, the (AI) learning system will compare all the specific similar product, e.g. mobile phone or computer, television, car etc. different kinds of specific products of global manufacturers to conclude the result is such as whether which brands will be the best manufacturers to let the consumer to buy the television or mobile phone or computer or car etc. different kinds of products. It can make more accurate judgement to compare general human's phone or questionnaire surveys investigation method, newspapers, television, radios, internet searches etc. different manufacturing news or data gathering channels to find which brands are the most worth confidence to consumers to choose to buy the specific product in the global consumption market.

On the manufacturers' beneficial hand, manufacturers can apply (AI) data gathering method to predict consumer emotion and buying behavioral changes more accurate. For example, the vehicle manufacturer, it plans to gather data to predict potential driving fast speed sport vehicle consumers' preferences trends in order to make the accurate judgement how to design its sport vehicles to attract many sport vehicle buyers who will choose to buy it's brand of any driving fast speed sport vehicles. It can attempt to apply (AI) intelligent learning system to gather global different brands of sport vehicle data concerns that all past driving fast speed sport vehicle buyer's preference of sport vehicle design. Then, the (AI) intelligent learning system gather global different brands of driving fast speed sport vehicle which had ever been purchased by the different country's driving fast speed sport vehicles consumers. After, it can compare divide the range of similar driving fast speed sport vehicle design and similar price to be different groups. The (AI) intelligent learning system can attempt to follow the past number of different brands of driving fast speed sport vehicle buyers to calculate how many driving fast speed sport vehicle buyers who choose to buy the brand of driving fast speed sport vehicle as well as it will analyze and make judgement to find whether the cheaper price reason attracts the different countries sport vehicle buyers choose to buy the brand of driving fast speed sport vehicle or the attractive design reason attracts the different countries sport vehicle buyers choose to buy the brand of sport vehicle or fast speed reason attracts the sport vehicle buyers choose to buy the brand of sport vehicle.

For example, although some brands of driving fast speed sport vehicle manufacturers' prices are very high, but they can still attract global many sport vehicle consumers to buy. Whether all sport vehicle's attractive design is the main factor to influence them to buy or whether it's fast speed is the main factor to influence them to buy or whether it's safe confidence it the main factor to influence them to buy or it's familiarity brand is the main factor to influence them to buy. (AI) intelligent learning system will attempt to make judgement and analysis to conclude whether the attractive design factor is the main factor to influence many sport vehicle consumers to choose to buy the brand of sport vehicles.

Otherwise, for another example, although some brands of driving fast speed sport vehicle manufacturer's prices are low, but they can not still attract many global many sport vehicle consumers to buy. Whether all vehicle's unattractive design is the main factor to influence them choose not to buy their fast speed driving sport vehicles or whether the unsafe factor is the main factor to influence them choose not to buy their fast speed driving sport vehicles or whether unfamiliarity brand is the main factor to influence many consumers choose not to buy their fast speeding sport vehicles.

Thus, when (AI) learning system had helped the fast speed sport vehicles manufacturer to gather all different brands of fast speed driving sport vehicle's past sale data and price data, design of different sport vehicle, e.g. color choice, method of style, comfortable chair styles and chair sizes and what kinds of steel material to manufacture the sport vehicles data and driving safe and accident occurrence data and the data concerns what reasons of the past complaint to brand of sport vehicle manufacturer from its sport vehicle buyers. Then, it can make more conclusion to give more accurate opinions whether which brands of fast speed driving sport vehicle manufacturer(s) whose sport vehicle design is the main factor to attract consumers choose to buy its any driving fast speed sport vehicle products really. Thus, it seems that it can make more accurate judgement to compare television survey, questionnaire survey to gather data concerns how to design the fast speed sport vehicle to attract consumers to choose to buy the sport vehicle manufacturer's planning sport vehicle products. I believe that (AI) learning system can help the sport vehicle manufacturer to make more accurate conclusion or judgement how to design its fast speed driving sport vehicles to attract it's consumers more easily.

(AI) tool predicts consumer immediate and expected emotion how to influence consumption decision

If (AI) tool can be confirmed to apply to predict consumer behavior, then I can conclude that it can be attempted to apply to predict what the factor(s) of the product itself can cause the consumer has positive or negative emotion, so the manufacturer can attempt to avoid the bad factors cause to bring negative emotion to influence the consumer chooses not to buy the product more easily, such as vehicle product case.

Economists refer to the consumption desirability is as " utility" and the product or service consumption decision making is arose

influenced by maximizing utility only. However, they neglect consumer individual immediate emotion change will also influence the consumer individual consumption decision consequently. Expected emotions are those that are anticipated to occur as a result of the outcomes associated

with different possible courses of action. For example, if a potential investor, were deciding whether to purchase a stock, who might imagine the disappointment who would feel if who ought it and it reduced its price. Otherwise,

whose emotion would experience , such as regret if it increased in price, but who does not buy it before the stock rise its price. However, I believe nowadays technology, in the future one day, (AI) tool can be attempted to assist consumer psychology profession or marketing research profession to assist them to find what are the bad factors to influence consumers choose not to buy any manufacturers' products. Then, when the manufacturer
can discover what are the bad factor(S) cause(S) consumers who do not choose to buy their products, then the manufacturer can raise whose product of consumption desirability or " utility" to raise whose product's consumption decision making is influenced by maximizing utility. Hence, (AI) tool will be possible to find what the bad factor(S) to cause consumers do not choose to buy the manufacturer's product in order to raise the product's utility to bring consumer positive emotion to choose to buy its product in possible. SO, (AI) tool will be one consumer psychological emotion prediction tool to assist any manufacturers
to help their products to build positive emotion to any consumers in possible.

The key feature of expected emotions is that they are experienced when the outcomes of a decision materialize, but not at the moment of choice, at the moment of choice, they are only feel about future emotion. Such as consumption case, if the consumer chose to buy the product or consume the service before it's price is increased. Then, the consumer will feel happy and it is worth to purchase or consumer the service as well as the consumer's expected emotion is positive before who decides to buy the product or consume the service, because who believes or feels the product or service's price will be raised in short term, e.g. after one month, one week. Thus, it means that if the consumer does not believe or
feel or predict the product or service's price either it will increase or decrease in short term, whose emotion will be negative, those negative emotion will influence who does not decide to buy the product or service, it is possible that who feel it is not worth to buy the product or consume the service immediately. He She will choose to consume the service or buy the product to wait it's price is decreased later. it seems that the consumer's positive or negative emotion will influence who decides to buy the product or consume the service later or earlier. Thus, it has close relationship between the consumer individual immediate purchase or consumption decision and positive emotion or negative emotion (either expected emotion or immediate emotion influences).

Consequently, if (AI) tool can help any manufacturers
to predict when its product price ought to be increased or decreased in order to attract consumer to choose to buy its product. Then, it can help any manufacturers to build positive expected emotion to attract consumers to choose to buy its product more easily. For example, when the (AI) tool can predict when the consumer expects the product price will fall down, then it can give ideas to the manufacturer to raise up the product price in the month, then it predicts many consumers expect the product price will fall down after six months. So, the product price will not be fall down after six months. So, many consumers will feel disappointment and they will choose to buy the product if the manufacturer
decided to raise the product price after six months. Then, the higher product price will cause many consumers worry about the product price will continue rise up, so they will prefer to choose to buy the product immediately after six

months because they afraid the product price will continue to rise up in the year. Then, I assume that (AI) tool has effort to predict when consumers feel the product will rise up or fall down, then it can give ideas to the manufacturer when to rise up or fall down the product price in order to attract or persuade many consumers choose to buy the product in different period in the year.

1.1 What does (AI) tool predict immediate emotion mean?

Psychologists indicate that immediate emotions, by contrast, are experienced at the moment of choice and fall into one of two categories. Integral emotion, like expected emotions, arise from thinking about the consequences of one's decision, but " integral emotion", unlike expected emotions are experienced at the moment of choice. Such as purchase stock case, the share buyer might experience immediate fear at the thought of the stock's losing value. " Incidental emotions" are also experienced at the moment of choice, such as a consumer predicts the product or service price whether it will be risen up or fallen down. If he/she feels the product or service price will fall down after next month and he/she will choose to buy the product or consume the service. But consequently, after next month, the product or service's price won't fall down absolutely.

Then, he/she will have incidental emotion to influence whom to choose whether he/she ought buy the product or consume the service, due to the product or service price is not still fall down. Otherwise, he/she is fear the product or service will not fall down in short term. Even, it will increase price later. Hence, whose incidental emotion will have possible to influence whom to choose to buy the product or consume the service after one month, if the product or service's price is still not increased absolutely. So, (AI) tool can be attempted to apply to predict when the product price ought need to be raised or fallen down in order to attract consumers to choose to buy the manufacturers' product in different period.

Economists indicate utility an individual consumption with an outcome might arise from a prediction of emotion: For example, a dinner eater might choose a higher utility to an Italian restaurant dinner than a French restaurant dinner because who anticipates being happier at the former, even the former's dinner price is higher than the French restaurant.

So, such as this restaurant dinner case, if one (AI) tool can assist the French restaurant owner to find what factor(S) cause(S) the dinner consumers do not choose to go to its restaurant to eat its food, e.g. high price factor, bad taste factor, bad wait service performance factor, bad cooker's cooking skill factor, poor advertisement promotion factor, poor familiar factor, poor location or poor eating environment etc. different factors. Then, the French restaurant owner can find methods to avoid the bad factor(S) cause(S) many dinner consumers do not choose to go to whose French restaurant to eat dinner more easily.

The question is that whether the positive emotion factor can influence the consumer changes whose mind to choose to consume the more expensive service or buy the more expensive product. To answer this question. it depends on whether the consumer has an imperfect understanding of whose own tastes or the consumer has a perfect understanding of whose own tastes to the product or the service.

It means the consumer will choose to buy the product or consume the service, even it's price is higher than other general similar products or services if who has a perfect understanding of whose own tastes to the product or service. Otherwise, who won't choose to buy the product or consume the service, due to it's price is higher than other general similar products or services if who has an imperfect understanding of whose own tastes to the product or service. So, it seems that the consumer's negative or positive emotion arise will be influenced by whose perfect or imperfect understanding of whose own tastes to the product or service factor.

It concludes that whether how much degree of the consumer's utility to the product or service. It is not the only one important factor to influence the consumer to choose to buy the product or consume the service. Otherwise, the consumer's imperfect or perfect understanding own tastes to the product or service factor will influence the consumer to arise positive or negative emotion to make final purchase or consumption decision immediately. It will be one more consumption influential factor to lead the consumer to make the final consumption decision making immediately. So, future (AI) tool ought to be innovate to own how to judge good taste or bad taste for any food in order to predict food consumers to choose to buy the food manufacturer's any foods more attractively.

(AI) tool technical innovation in cruise tourism
immediate positive emotion influence to
cruise travelling consumers

Can apply (AI) tool to cause positive emotion to cruise tourism consumers? Cruise tourism industry is the most influential emotion industry example to influence cruise travelling consumers' travelling entertainment choice. I shall indicate some evidences how it's innovation will influence cruise travelling consumers' emotion to be changed to positive from negative immediately as well as to prove how the cruise traveler higher utility feeling to the cruise tourism provider is not the main factor to influence whom to choose the cruise provider to consume whose cruise journey service arrangement.

Nowadays, cruising has become one of the fastest growing sectors within tourism, cruise service providers need have themselves unique different entertainment service arrangement to satisfy every cruise travelling consumer individual needs in order to attract every one to choose whose cruise arrangement easily, e.g. meals, activities, entertainment and varied destinations create one-stop holiday shop, reasonable competitive ticket fare. Hence, it seems it is one exciting emotion industry. If the cruise service provider can bring positive emotion to influence many cruise travelling consumers immediately. The, even it change higher service fare to compare other similar cruise service providers. I believe it won't influence them to choose other similar cruise service providers if it can often bring immediate positive emotion to its cruise clients during they are staying in its cruises or during they have left its cruises, but they will often remember or won't forget to enjoy their cruise service provider's happing time forever. Hence, if (AI) tool can be attempted to help cruise entertainment providers to arrange different cruise journeys for varied destinations , to arrange different entertainment facilities, to arrange the different taste food to satisfy different countries age cruise consumers' needs. Then, the (AI) tool will assist the cruise providers to bring positive emotion to let every different countries age cruise consumers to feel satisfactory in order to choose to the cruise providers' cruise entertainment service more attractively.

2.1 How can apply (AI) tool to predict cruise service providers bring positive emotion to their clients?

Future, (AI) tool can help any cruise providers to design these kinds of any one entertainment service arrangement to satisfy the cruise provider's customers' needs.

There are different special interests cruising , such as wellness at sea, freighter cruises, river cruises. It has increased the attractiveness of cruising: Romance is for lover cruise traveler target, luxury is for rich cruise traveler target, exotica is for enjoyment exciting feeling traveler target. So, every kind of cruise traveler target will have different kind of cruise entertainment service to satisfy their needs. If the cruise service provider can provide the right and attractive cruise entertainment service to satisfy the specific cruise target. Then, it will bring the positive emotion to the specific cruise target consumers more easily.

Cruise travel was shaped for mass tourism. Prices have been very differently segmented. There are basically four types of markets (Biederman, 2008):

- Contemporary market: On board fun and amenities are playing important role and destinations have secondary importance.
- Premium market: This category is more expensive than the contemporary category and where the destination has same importance as on board amenities.
- Luxury market: It was once dominant type of cruise tourism, but now it has only a small portion of the industry. Generally, it is the most expensive cruise category and usually it takes longer than average cruise days.
- Adventure/exploration: It refers relatively long cruises with special and exotic places where the destination is the main purpose of the trip.
- European cruise travel: Duration takes more five days than worth American travel duration. There is a tendency on European market during the years that duration of travel is getting shorter. This short demand of is explained with the strong demand of customers (Hensen, 2003). Beside this, it is most likely that cruise companies try to convince tourists with short haul travels instead of long term cruise trips for more expenditure.

Thus, I believe that even, the cruise service provider charges higher ticket which won't influence cruise consumers who do not choose its entertainment service on its cruises. If it can arrange the attractive cruise entertainment facilities and destination journey arrangement, staying days arrangement to satisfy different specific cruise target market needs absolutely in order to bring whose emotion to be positive to it's service provision. Then, the cruise service provider will attract many potential cruise clients to choose its cruise service absolutely. Otherwise, if it only bring negative emotion to its cruise clients, it will not attract many potential cruise clients to choose it or loses its old cruise clients, even, its cruise ticket price is needed to decreased in order to raise competitive effort.

In conclusion, I believe that future (AI) tools need to learn how to bring cruise consumers to arise individual immediate or expected positive emotion, this positive emotion consideration is more important to compare to how to reduce cruise ticket price in order to attract cruise clients in global cruise competitive cruise industry.

2.2 Differentiation through the characteristics of cruising route method from (AI) tool route judgement

Future, (AI) tool can attempt to help any cruise entertainment service providers to judge how to design different route to attract different countries age cruise clients' choices to satisfy their cruise journey entertainment needs. The determinants of the cruising route's characteristics (functional, social, and emotion) is important factor to influence the cruise service provider's success. Cruising product is no longer selected primarily for the cruising service, but for the content of cruising route. So, the cruising route will influence the cruise consumer individual emotion, because it is the main service need for every cruise consumer.

The approach called the " land sea cruising in product development" is increasingly becoming an area of interest, e.g. determining the direction of the effects of the individual cruising route characteristics on service value's perception , and providing an evaluation model of the route's perception , and indicating significance variables of attraction.

The questions that cruise planners need to know: How does each of the identified determinants affect the overall perceived value of the cruise route? How the overall perceived value of the cruise route affects customer behavior intentions?

Because different routes factor will influence cruise consumer individual emotion changing seriously. It means the ship has become only a tool, when the offered route whose attractiveness highly influences the impression of the guests has become crucial.

Consumer behavior in cruising segment includes all the activities and influences in the selection of the specific cruise route. There activities result in decisions and actions related to a defined price, selection and reselection of cruising company (Cannot, Brink and Brijball, 2006).

2.3 How to apply (AI) tool to arrange cruise route planning have close relationship to influence cruise consumer emotion?

Firstly, use value of cruising routes is based on the subjective experience, and shows how individuals assess the route during, or immediately after sailing. It is affiliated with the benefits that cruising guest realize by choosing a route , and it is subjective because it depends on the individual assessment (photo taken on the route for one guest presents just a family souvenir, and for professional photographers are embodied financial capital).

Secondly, the utilitarian value is also subjective-oriented and is tied on the point where the inner and us ability of cruising routes are compared with the sacrifice of the client (money and time). Finally, the value is considered as the outcome of the comparison of scarifies and personal benefits, which is resulted in essentially utilitarian nature.

Hence, route design is the main value of cruising tourism and it is primarily determined and analyzed from the aspect of observed customers. Otherwise, the cruise is only one tool to be caught for the cruise passengers, whether the cruise can let whom to sleep comfortable , providing what kind of food to them to eat, what kind of entertainment facilities are provided to them to play, these issues are not more important to compare how to design route to bring them to travel to anywhere to enjoy in this cruise journey factor. Because how to design the route factor can bring each cruise passenger to influence them to feel either negative or positive emotion directly. The whole route journey planning is the most influential factor to influence the cruise passengers to feel whether they ought choose it's service again or not in the future.

(AI) tool judges the difference between utility factor and emotion to influence consumer decision making

In economic utility or immediate (expected) emotion aspects, whether which is more influential to excite consumption. To analyze whether it is economic utility or immediate (expected) emotion more influential to excite consumption. It depends on the consumer individual consumption choice is in which situations. For example, if the industry's general consumer individual consumption decision is concentrate on emotion influential aspect, such as cruise entertainment industry, hospital care service industry, theme park entertainment industry, movie watching entertainment industry etc. Above all these industries have same nature, it is service. So, it seems that service industry's main influential factor is immediate (expected) emotion influence, it is not economic utility influence. Otherwise, product sale industry's main influential factor is utility.

3.1 (AI) judges consumer utility factor

For this toy choice situation example, parent choose to buy one toy to give whose child to play. They usually considerate which kind of toy is attractive to their child whom like to play. In many different kinds of toys choice, if the child likes to choose the kind of toy to play. After the child's parents had purchased the kind of toy to let whose child to play one period time, e.g. six month. Then, when the child feel that who has need to buy another new toy to play, due to he/she feels bored to play this toy. So, it seems that the child feels this toy has less utility or it's utility is decreased. So, he/she expects whose parent can buy another new kind of toy to let whom to play. It also implies that it is not emotion factor to influence the child to feel boredom and unfunny to play this kind of old toy after six months. It is the product's utility factor which can not attract the child to play it any more. So, this old toy's utility is decreased when this child spends six months to play it. This toy's value is only six month utility to this child to play. Otherwise, if this kind of toy is bought by another parent. It is possible that the another child like to play this kind of toy one year or more. So, it's utility to another child is one year or more period. So, product's utility period is difference, it depends on how long time of the user's satisfactory time.

As this toy case, the child's decision will influence whose parent choose which kind of toy to buy to whom to play. Usually toy price is not difference too much. Parent won't consider when the toy price will be increase or will be decreased to influence their emotion to decide not to buy the toy immediately. So, when the child like to play the kind of toy, even the product's price is more than other kind of toys, and the parent feel it is possible that the kind of toy's price will be fallen down later. They will still choose to buy the kind of toy to let their child to play, they won't be influenced not to buy this kind product by later cheap price factor. So, immediate emotion is not the main factor to influence this parent does not choose buy this toy at this moment. Otherwise, utility factor will influence when the parent will buy another new kind of toy to provide to this child to play. If the child enjoy to play it only three months, after he/she will feel bore and he/she will tell whose parent to buy another new kind of toy to let whom play when the fourth month is beginning. So, it implies that if the kind of toy product can have more attractive utility time, then it can attract many parent to choose to buy it among different kind of toys. Thus, when this kind of toy's utility time is longer time. Then, it is possible that it can influence many parents choose to buy it's different style or design of similar kind of toys to let their children to play. In general, when many parents accept to buy this kind of different style or design of similar toys to give their children to play. Due to it's popular long time utility factor, it will influence children like to play it longer time to compare other kind of toys. Consequently, it will influence parents do not need often spend too much money to buy other kinds of toys to give their children to play. So, longer time utility factor to the product can attract many consumers to choose to buy the kind of product to compare lesser time utility factor to the product. Hence, it proves the explanation why utility factor is the main influential factor to influence the consumer choose to buy the product.

3.2 (AI) judgement tool of Medical care and utility case

Medical care is an input in producing health, it is subject to law of diminishing marginal productivity. Health yields utility to the consumer. It is subject to law of diminishing marginal utility. It bring this question: Does either the patient's emotion or the medical care service or medical care product utility which one can influence the patient's hospital choice more?

To answer this question: We need to know medical care is one kind of nursing care service in hospitals or clinics and medical care product is one kind of medical care product sale from merchants, e.g. medicine or medical equipment. So, in medical industry which has different kind of medical care services to provide to patients in hospitals or clinics as well as which has different kind of medical care products sale, e.g. medicine or wheelchairs, heart health measurement equipment etc. different medical care products in medical health industry.

In medical care aspect, it is one kind of any medical care service to patients from hospitals or clinics. So, medical care is an input in producing health service to patients from hospitals or clinics. When the patient is admitting to hospital or clinic, who needs to see doctor and the doctor need to give the right medicine to the patient to eat to ill whose illness. Even, if the doctor feels the patient whom needs to live hospital for one time period. Then, the hospital nurses must need to take care the patient during he/she is living in the hospital period. Consequently, if the patent can be health in short time, e.g. within one week leaving time, then he/she will be shortened time to leave the hospital in next weak. Otherwise, if the patent can not be health in short time, e.g. within on week, then he/she needs to live the hospital more than one week, even, one month, three months or more. So, the staying hospital time will influence the patent's emotion to feel whether the doctor's effort. If he/she needs to live the hospital long time, he/she will bring negative emotion to feel the doctor's medical effort is not good. The doctor's medical effort can not achieve or satisfy whose expected emotion during whose staying hospital time.

Thus, medical care is an input service in producing health, it is subject to law of diminishing marginal productivity. When the patient does not need to live the hospital longer time, the patient will feel more satisfactory to the hospital's doctor and nurses' care effort as well as the patient can give less money to spend the expenditure to live the hospital. So, the law of diminishing marginal health productivity will explain the hospital will shorten time to the staying days of the hospital to the patent as well as the patient's care expenditure will be decreased when he/she only needs to live to the hospital in short time. Otherwise, the patient needs to live longer time in the hospital, it means that the diminishing marginal health productivity to the hospital, the patient's staying hospital days will be increased and the patient's medical expenditure to the hospital will also be increased. It will bring negative emotion to patient and why this negative emotion factor will influence the patient would choose another hospital to live or find other doctors to see if he/she felt illness in future one day.

In medical care product aspect, health yields utility to the consumer. It is subject to law of diminishing marginal utility. Because patient needs to buy different kind of medical equipment to use or medicine to eat to attempt to cure whose illnesses. So, if the patent can choose the right medicine to eat from the doctor's recommendation or if the patent can choose the right medical equipment to use from the doctor's recommendation. When the patient buy less number of medicine to eat, then he/she can be health or when he/she buy the medical equipment to use, then he/she can be health. Then, he/she will spend less money to buy medicine to eat or medical equipment to use and he/she can be health in short time. Then, the medical consumer will feel the medicine or medical equipment has good utility to satisfy whose medical needs. So, good medicine and good medical equipment can only need short time and less money to let the medical patient to be health.

3.3 Immediate (expected) emotion factor

As cruise entertainment case, every cruise journey must provide fixed stay days on the cruise to let every cruise passenger to play to every cruise journey. So, cruise passenger can not change or extend whose fixed stay day choice in every cruise journey, when they had caught the cruise to go to sea on the day. Is implies that cruise entertainment has none longer time utility factor which can influence each cruise passenger's choice to each different design of cruise journey arrangement.

If the cruise passenger feels very satisfactory and enjoyable to the last time of specific cruise journey arrangement, e.g. five days and four nights New Zealand and Australia cruise journey. Due to this cruise journey can bring positive emotion to let the cruise passenger to let whom feel that he/she can not forget or remember this happy cruise journey forever.

It is possible that this time happy five days and four nights New Zealand and Australia cruise journey will bring positive emotion to influence this cruise passenger to choose to find this cruise service provider to help whom to arrange this same cruise journey or another similar cruise journey again after one month, or three month, or six month or one year or more. Due to this cruse passenger felt this cruise service providers' cruise journey design arrangement can satisfy whose needs and it can achieve whose expected emotion to be positive. So, this cruise entertainment industry must be immediate (expected) emotion influential factor more than time utility factor to influence the cruise consumer's cruise service provider and cruise journey choices.

In conclusion, it is not only utility factor can influence consumption decision. It is emotion factor can also influence consumption decision. It depends on situations whether the consumer is choosing to buy one product or consume one service. If the consumer is choosing to buy one product, how long time of the product's utility factor which will influence the consumer choose to buy which product. If the consumer feels the product can give longer utility time among other similar products, then he/she will have more chance to choose to buy the product. Otherwise, if the consumer feels the product can give lesser utility time among other similar products, then he/she will have less chance to choose to buy the product. If the consumer is choosing to consume one service, emotion factor will influence the consumer choose to find which service provider to consume the same service or similar service. If the service provider can provide excellent service to the consumer, then it will bring positive immediate or expected emotion to whom and it can attract the consumer to choose the service provider again. Otherwise, if the service provider can not provide excellent service to the consumer, then will bring negative immediate or expected emotion to whom and it can not attract the consumer to choose the service provider again.

Why and how (AI) judgement tool can judgement what utility factors are to influence consumer emotion

In emotion and utility both aspects, they include these situations. I shall explain how and why emotion and utility factors can influence consumption behaviors in these different situations as below:

(1) In the first situation is brand factor, the brand image, product quality, product knowledge , attitude and

(2) brand loyalty intangible factor will attract the consumer individual purchase.

For example, luxury

products, e.g. luxury fashion brands of clothing. Brands like Zara from Spain and H&M from Seweden began to produce catwalk-style fashion at low cost offering consumers of luxury fashion alternatives at low prices.

Nowadays, the luxury fashion sector is the fourth largest revenue generator in France, and one of the most remarkable sectors in Italy, Spain , the USA and the potential markets of China, Russia and India. The luxury industry has increased having a huge youth in demand. The luxury consumer have much choice in products, shopping channels and pricing of luxury products. It has possible relationship of age, gender, income and other demographic factors with purchasing intentions to influence the rational and emotional buying behavior regarding luxury fashion products.

(2) The second situation concerns the decision-making of make or female consumers are possible experience an emotional desires and cognitive (reasoning) mind in purchasing choice process. Their emotion includes negative or positive buying emotion and mood management and cognitive process components include cognitive deliberation, planning buying with the exception regard for the future.

University had been using analysis of variances tests, male and female students were found significantly different with respect

affective process components including positive buying emotion, and mood management and cognitive process components include planning buying.

Significant differences were also found between the following product categories: shirts/sweaters, skirts, coats, underwear, accessories, shoes, electronic hardware, computer software, music , CD or DVDs, sports, memorabilia, health /beauty products and magazines/books for pleasure reading. No differences were found in regard to suits/ business wear and entertainments.

The investigate proved that some products will have different emotion influence to cause female or male students whose final consumption decision to buy the kind of product. So, the difference od male and female students will have emotion influence to make purchase decision to buy the product in consuming choice process.

(3) The third situation concerns search advertising factor, e.g. online search to influence consumption behavior. Advertising is possible one method to persuade the consumer to choose to buy the product, even the consumer does not know the product exists. For example, proper cloth, a company based in New York, has a site on the social networking site Facebook.

Whenever the company posts a new photos of its clothes, all its face book " fans" automatically receive the information on their own face book pages. "We want to hear what our customers have to say." It seems online advertisement is a potential promotion method to promote any new attractive products to sell to let publicity to know to buy. Internet is one popular communication tool to be used by youth today. So, when one company can have one website to let any youth to find and enter to the website to discover any new things easily. It will cause many consumption chances to let online potential clients to attempt to choose any products to make purchase decision from online advertising tool easily.

How does the role of advertising influence the purchase decision process? Needs and motivations are the starting points of purchase decisions. In fact, advertising is a communication of photo image, sound image, and word advertisement image channel to persuade consumers to choose to buy the brand of product or consume the brand of service between the merchant and its consumers from television, radio, newspapers, magazine, movie etc. channels.

Why does advertisement influence consumer
choice? For this case example, when one buyer waits until more information is gathered before making a decision. The time, two types of cost are involved. First, there are psychological opportunity costs experienced by consumers who are deprived of the product who need and are consequently in a state of psychological tension.

As time elapses, this psychological tension becomes more frustration. Second, buyers experience costs with the information-gathering efforts. They must invest time and energy to visit several retailers, seek out and read advertisements, or inquire for other opinions about the best product to buy.

These delayed decision costs considerably increase as time elapses. The buyer must seek information until it is felt that a search for additional information will bring about more costs than benefits. So, an advertisement is reaching a potential buyer when who is seeking information will have a greater impact, since the buyer is spending time and effort needed to seek out this information himself and he is less likely to find other competing and advertisements to obtain the additional information.

In general, buyers are generally more responsive to different brand advertisements, when they are seeking information on these brands. This is why the becomes a choice target for the advertiser provided the advertiser can identify and locate them. Thus, a client has interested and is in an information-gathering stage is asked.

Then, the advertiser takes advantage of the consumer's having identified him or herself to send a series of informative and persuasive messages or to send a salesperson who will try to conclude a sale. Thus, advertisement gives a chance to let consumers to gather information to choose the best product to buy or the most excellent service to consume.

What of situations do merchants need advertisements promotion? The short purchase cycle markets are characterized by routine

purchase decision processes or by limited problem solving when a new brand is introduced on the market, e.g. coffee, bread , sugar, soft drinks, canned vegetables and household and beauty care products fall into this category. Another irregular purchase cycle markets are characterized by products that are purchased more or less regularly , e.g. cookies, cake mixes, wines, food products. Finally, long or unpredictable purchase cycle markets include all durable products, such as cars, household appliances and furniture (products from which occasions of purchase can't be predicted, which most consumers buy only occasionally). Hence, these kinds of products ought need advertisement promotion specially, due to advertisement can build brand image to let consumers to know. Especially , it is a new brand of product. In conclusion, advertisement will be one good channel to let new product to introduce its brand to let clients to remember in minds heart.

Consequently, future (AI) tool needs to learn how to design different kinds of advertisement to attract consumer attention to the product, needs to learn how to find what the bad factor(S) which cause(S) many consumers do not choose to buy the product, needs to learn when the product price needs to be raised up or fallen down in order to bring consumers' positive emotion to choose to buy the product immediately. SO, if (AI) tool can be invented to own itself effort to design different kinds of methods to predict consumer behavior in order to bring their positive emotion to the product successfully, then the manufacturer can earn positive consumer emotion advantage from the (AI) tool assistance for long term benefit to its product.

How can artificial intelligent tools predict travelling consumer behavior in airline and air agent travelling market ?

I believe that applying (AI) big data tool to predict vehicle buyer consumption choice behavior, it is similar to predict traveler consumption choice behavior. In this chapter, I shall indicate how to apply (AI) big data gathering tool to predict vehicle buyer consumption choice behavior. Then, I shall its what its similar points to be applied to predict traveler consumption choice behavior.

Nowadays, many vehicle manufacturers hope their vehicles can attract to vehicle buyers to choose to buy their vehicles. However, there are many different brands of vehicles to provide to them to choose, so the vehicle market competition is very serious.

How to judge their different kinds of vehicle price which is reasonable acceptance to attract vehicle buyers to choose to buy the brand of vehicle manufacturers' any kinds of vehicles, e.g. fast speed sport style vehicles, comfortable and slow speed common cars, for four passengers common small size or more than four passengers common large car size?

How to evaluate the vehicle prices issue is important factor to influence vehicle buyers' choices. Either if the brand of vehicle price is too high to compare other brands of similar vehicle price, it will influence many vehicle buyers choose to buy other brands' vehicles or if the brand of vehicle price is too low, it will influence vehicle buyers feel this brand's vehicle machine quality or safe driving level or manufacturing steel material or speed or not comfortable sitting etc. different factors is worse to compare to other vehicle brands' similar vehicle products.

Thus, if the brand of vehicle manufacturers can predict how to design vehicles which can attract many vehicle buyers to choose to buy whose any vehicle products. What are future vehicle buyers' favorable vehicle styles? Then, the vehicle manufacturer can concentrate on manufacturing the kind style of vehicle products to sell already. It will reduce its vehicle manufacturing investment risk.

How to apply (AI) tools to predict vehicle buyers' behavioral consumption model? Whether artificial intelligent tools can predict automotive buyers' behavioral consumption model and predict future vehicle design trend. In fact, automotive brands and dealerships are facing an increasingly competition when attempting to manually gathering the vast quantities of data required to create customer focused programs that increase retention, ultimately new sales and service automotive business.

Building a based on that client's intrinsic needs and interests to any kinds of automotive vehicles at any given time. This is especially true in the automotive industry where the time span between purchases is measured in years. Because vehicle buyers would not like often to change their old vehicle to another new one. So, their decisions to buying another new vehicle, the time is usually after one year, even longer time. Hence, it seems any vehicles won't be frequent consumption products to the owned at least one vehicle family consumers (vehicle buyers). It implies that why vehicle manufacturers ought need to spend time to predict future vehicle buyer design choice for whole year vehicle buyer number growth because they won't often change preferable vehicle design to change another new vehicle more easily.

Hence, how to predict vehicle consumers' taste or preferable which styles of vehicle choices issues is very important. If the vehicle manufacturers can not manufacture any attractive vehicles to sell easily in this year. Then, it will lose time, money in this year because it won't know when the owned least one vehicle users or non-owned any vehicle users who will decide to buy one new vehicle or change another new vehicle ensure. The different brand vehicle dealers will possible wait more than one year to attract them to buy their vehicles if their styles are not attractive to compare other brands of vehicle competitors.

However, artificial intelligence and machine learning can help any vehicle manufacturers to find solution to solve patterns in highly to solve patterns in highly complex data-sets that are beyond the capability of a human brain, and then building and automatically acting on the customer insights it generates.

Given the automotive customer need for individualized communications, this technology is positioned to become a critical component of any successful vehicle retailer's domestic or/and overseas vehicle markets. How can vehicle manufacturers and retailers use (AI) to enhance their vehicle marketing campaigns? How will (AI) affect their vehicle sale marketing strategy? What criteria would they use when selecting on (AI) solution?

Vehicle consumers today are able to quickly access different brands of vehicle information, research vehicle products and reviews, negotiate prices and compare one vehicle brand or retailer to another resulting of the brands of vehicle customers. At the same time, the rise of " big -data mining", wearable devices that track user's every move and preference and greater contextualization in advertising and social media has resulted in consumer expectations of individualized. Thus, it seems that (AI) tools can be used to gather " big-data" and then they can make human's mind to analyze how to design kinds of vehicles to satisfy vehicle buyers' needs.

As automotive vehicle marketers can apply (AI) tools to achieve messaging strategies to meet the needs of this new generation of informed vehicle consumers, using data from a variety of sources to move from a variety of sources to move from mass- messaging to more personalized messages aimed at particular vehicle buyer segments, e.g. fast speed sport vehicle buyer segment, slow speed comfortable small size or large size of buyer segment. However, when 90% of vehicle marketers believe having a single vehicle buyer view is important, only 6% have achieved it.

However, one of the main issues vehicle marketers are facing the lack of capacity to efficiently sift through and analyze the massive vehicle buyer amounts of data required to create vehicle buyer individualized vehicle customer experiences easily. This is especially difficult for automotive dealers, the long periods between purchase cycles, and the highly considered nature of the vehicle purchase means that each vehicle dealer needs to not only track a large number of potential vehicle customers for an extremely long period of time, but each of those vehicle customers will generate a huge amount of different kinds of vehicle behavioral consumption data as they research their next vehicle purchase. However, by choosing the right (AI) technological tools and programs , vehicle dealers can solve this big data gathering challenge into a major advantage.

For Forrester vehicle brand example, vehicle consumers have more power over the Forrester vehicle brand's reputation than ever before. Mayne, L. (2014) indicated that Forrester calls this new (AI) tools is the " age of the vehicle customer", a 20 year business cycle in which the most successful vehicle enterprises will reinvent themselves to systematically understand and serve increasingly powerful vehicle consumers. To win in this new age, Forrester declares companies must become vehicle customer obsessed and the only sustainable competitive advantage is knowledge and engagement with customers, such as (AI) gathering data knowledge.

Thus, the biggest challenge vehicle businesses currently face is not the collection of a large quantity of vehicle consumer data, but what to do with that data once they have it. Even at a large vehicle data research firm, the data sets are often too big for a single analyze, or even a team of analysts to sort through and draw conclusion from. However, enter artificial intelligence and machine learning , an efficient technology solution that can continuously find patterns in highly complex data sets that are way beyond the capacity of a human brain and then automatic drive action based on the customer insights is generated.

What is (AI) machine learning tool? Machine learning is a type of (AI) that learns from data and is not explicitly program. Think Amazon, face book. Machine learning serves up relevant content based on an individual vehicle purchase behavior and experiences. More simply, machine learning is a computer program that can learn relationships between data, subject those learnings to errors functions, and then learn from its errors. The program in effect, trains itself.

Lee, T. (2016) explained that "Thus, (AI) tools can learn deep a more advanced branch of machine learning inspired by how our brain's nervous function, has also been found to be especial effective in identifying patterns from data."

When this way sound is complicated from a vehicle dealer perspective, the implementation of a marketing program driven by artificial intelligence can take care of these tasks in an automatic vehicle fashion with little to no manual intervention required from the staff at time vehicle stores.

In practice at a vehicle dealership, the program will continue track vehicle customer behavior online, merging that data with any offline source (like CRM or DMS data) and then analyze this aggregated vehicle buyer data set to predict what vehicle customer may be shopping for and what information they might like to relevance from different kinds style of vehicle design photos.

5.1 Why does travelling market seem to similar to vehicle market which can apply (AI) learning tool to predict travellingconsumer behaviors?

Artificial intelligence refers to complex in vehicle market and travelling entertainment market which is very seem to be applied to predict consumer behaviors.

(AI) machine learning that posses the same characteristics of human intelligence and that have all our sense, all our reason and think just like human vehicle buyer who prefer vehicle purchase choice or travelling consumer who prefer travelling package or travelling destination and airline choice. Besides, machine learning is the practice of using algorithms to collect and examine data, learn from it, and then make a determination or prediction about something in the world.

So, it can be attempted to gather data concerns that travelling consumer past travelling destination choice and air ticket price choice and different travelling package, e.g. high, middle, or low class hotel and foods supply and entertainment places choice in their past travelling journeys.

The machine is " trained" using large amounts of data and algorithms that give it the ability to learn how to automatically perform a task with increasing accuracy. Otherwise, deep learning is primarily based on artificial neural networks inspired by our understanding of the biology of human's brains.

Thus, (AI) big data can gather all these past traveler consumption behavioral choice data to make reference to analyze whether how many travelers will choose to go to the specific travelling destination in any time by the past traveler number record to different travelling destinations, then it can gather the past air ticket sale price to different destinations and past travelling package design to different destinations in order to analyze whether it is the cheap airline ticket price factor or attractive travelling package factor or attractive travelling entertainment etc. in order to predict which factor is the most potential influential factor to they choose to go to the destination to travel in different time within one year. Then, traveler agent or airline can collect these big data to judge how to design their package to attract travelers to go to anywhere to travel or what the main factor influence most of them to choose to visit the destination to travel.

For example, travel agents or airlines can apply "Deep learning" breaks down tasks in ways that enables machines to assist them to predict when travelling consumer choice will be changed and why their travelling choice will change and how their travelling choice will change with increasingly complex tasks.

So, such as why (AI) technology can be applied to predict how travelling consumer behavior changes to bring to judge whether anywhere will be many travelling consumers who will prefer to choose travelling hot destinations next year or next month.

Then, travel agents and airlines can gather overall past travelling consumer data to analyze and conclude the more accurate prediction of different travelling destinations to the number of traveler. Then, they can choose how much air ticket price is more reasonable to charge to the travelling destination or how to design the travelling package which can bring more attractive to the prediction number of different travelling destination travelers in order to achieve to raise the different travelling destination number next year.

Thus, (AI) big data machine learning can help airlines or travel agents to solve how to design any attractive travelling package challenge. A travelling package is both one of the most important and carefully considered travelling entertainment consumption the majority of travelling people will ever make in their lifetime at least one travelling time.

It is also a prediction how travelling package will be designed that tends to be fundamentally tied to a travelling person's travelling destination choice identify and travelling package view of themselves. As the same time, travelling consumers' travelling choice changing lifestyles result in changing travelling destination needs, e.g. the country's young travelers can choose to change non-extreme exciting travelling entertainment package from past extreme exciting travelling entertainment package. Due to personal feeling factor in general. However, I believe that (AI) big data can also be attempted to predict when the country's young travelers will choose to change non-extreme exciting travelling behavior.

It is similar to automotive dealers need to remember that vehicle customers and prospects are individual human beings with risk, complex and ever-changing lives factors, these factors will influence every vehicle consumer why who feels has vehicle purchase need, and how who choose to buy the first vehicle if who decided to buy the first vehicle.

It seems that travelling agents or airlines need to remember that travelling consumers and different features or designs are very traveler beings with risk, complex and ever-travelling package attitude personal changing factors in different travel season, these factor will influence every individual traveler why who feels has travel entertainment need, and how who choose to buy different feature or design travelling package if who decide to travel.

The (AI) big data technological travelling customer behavioral prediction tool seems to be the best travelling behavioral prediction tool in the world are those that know every one of different country's traveler need. Their likes and dislikes which style of travelling package, preferences and travel destination changing tastes to travelling destination choices.

The capacity of the human brain, however, limits us from achieving these different type of travel package sales. In this competitive travelling destination choice entertainment environment, (AI) big data machine learning enables platforms to assist the air ticket and travel package sales team by tracking the travelling consumer behaviors of each travelling customer, learning and memorizing their preferences and predicting their future travelling destination choice and travelling package design needs.

Finally, I recommend that for a travel agent or airline travelling marketing platform to make their travelling customer engagement efficient and fully-functional, I should be able to: applying (AI) tools to track every travelling customer behavior across the web, connecting to a society of data sources, CRM, DMS, third-party, web travelling brands, social traveler email, click etc., aggregating and accurately cross-reference data from a variety of sources, leveraging this data to drive insights on a mass scale, as well as on an individualized basis, driving actions and automatically direct travelling customer engagement via multiple channels based on where each customer is in their travelling individual lifecycle.

5.2 Why is (AI) big data gathering tool better than psychological and survey methods to predict traveler individual travel choice behavior?

Prediction travel behavioral consumption from psychology and survey methods.

How to predict travel consumption? It is one question to any travel agents concern to use what methods which can predict how many numbers of travelers where who will choose to go to travel more accurately. I think that who can consider how to predict travel behavioral consumption from psychology and survey travel choice prediction method, but it is better to apply (AI) big data gathering method to predict travel consumer's destination choice more accurate. The reason is as below:

The first reason is that traveller individual travel psychological desire is difficult to predict accurate more than (AI) big data gathering method, it is due that the data is past traveler's destination choice and travel package and ticket price actual data from (AI) big data gathering method. Otherwise, survey investigation is only traveler psychological thinking method. It lacks enough past actual traveler data gathering.

The second reason is that on the weakness of traveler individual psychological thinking view of survey investigation. It has evidence to support the relationship between self-identify threat and resistance to change travel behavior to any travelers, controlling for whose past travelling behavior, resistance to change if a psychological phenomenon of long standing interest in many applied branches of psychology.

Past travelling behavior has been acknowledged as a predictor of future action. Such as travelling behavior that is experienced as successful is likely to be repeated and may lead to habitual patterns. Some psychologists differentiate habit between two concepts, such as goal oriented and automatic oriented both. Although repeated past travelling behavior is addition goal oriented and automatic oriented. Further non-deliberative nature of habit may make appeals to judge and to predict future individual traveler's behavior accurately.

However, repeated one traveler will choose the destination to repeat to travel without a necessary constraint of goal orientation and automatic oriented both. So, it seems that psychological factor can influence any individual traveler why and how who choose to decide to repeat to choose the destination to travel.

So, survey investigation is only the traveler's thinking to answer the travel firm. It is not sure that the traveler's past travel experience is real answer. Otherwise, (AI) big data gathering method is computer gathering method which gather past traveler consumption actual data to analyze and conclude future traveler possible repeated travel destination choice and travel package choice more accurate.

The third reason is that on the strength of (AI) big data gathering method computer statistic view to predict future traveller consumer's destination and travel package choice. It is structural equation modeling is an extremely flexible linear-in-parameters multivariate statistical modeling technique. It has been used in modeling travel behavior and values since about 1980 year. It is a software method to handle a large number of variables, as well as unobserved variables specified as linear combinations (weighted averages) of the observed variable.

5.3 Can (AI) big data gather data to predict when climate will change to influence poor travelling behaviours?

(AI) big data tool can predict the flexibility of human travelling behavioral change is at least the result of one such mechanism, our ability to travel mentally in time and entertain potential future. Understanding of the impacts is holidays, particularly those involving travel.

Using focus groups research to explores tourists' awareness of the impacts of travel own climate change, examines the extent to which climate change features in holiday travel decisions and identifies some of the barriers to the adoption of less carbon intensive tourism practices.

The findings suggest many tourists don't consider climate change when planning their holidays. The failure of tourists to engage with the climate change to impact of holidays, combined with significant barriers to behavioral change, presents a considerable challenge in the tourism industry. In the future, computer (AI) big data tool can attempt to predict when the country's climate change to influence travelers to choose to go to the country to travel, e.g. next month or next half year or next year hot travelling destinations.

Tourism is a highly energy intensive industry and has only recently attracted attention as an important contributions to climate change through greenhouse gas emissions. It has been estimated that tourism contributes 5% of global carbon dioxide emissions. There have been a number of potential changes proposed for reducing the impact of air travel on climate change. These include technological changes, market based changes and behavioral changes.

However, the role that climate change plays in the holiday and travel decisions of global tourists. How the global tourists of the impacts travel has on climate change to establish the extent to which climate change, considerations features in holiday travel decision making processes and to investigate the major barriers to global tourists adopting less carbon intensive travel practices.

It will bring this question: Will (AI) predict tourists aware the impacts that their holidays and travel have on climate changes to influence their travelling decision?

When, it comes to understand individual traveler's behavioral change, wide range of conceptual theories have been developed, utilizing various social, psychological, subjective and objective variables in order to model travel consumption behavior. These theories of travel behavioral change operate at a number of different levels, including the individual level, the interpersonal level and community level. Whether pro-environmental behavior can be used to predict travel consumption behavior in a climate change. However, the question of what determines pro-environmental behavior in such a complex one that it can not be visualized through one single framework or diagram.

Despite the potentially high risk scenario for the tourism industry and the global environment, the tourism and climate change ought have close relationship.

However, (AI) big data tool can be applied to find what factors to influence the time of travelers' travelling choices. What are the important factors and variables which can limit tourism? e.g. money, time, family problem, extreme hot or cold weather change, air ticket price, journey attraction etc. variable factors.

Mention of holidays and travel were deliberately avoided in the recruitment process, so as not to create a connection factor to influence traveler's individual mind. However, the dismissal of alternative transportation modes can be conceived as either a structural barrier, in the sense that flying is perhaps the only realistic option to reach long-haul holiday destination, or a perceived behavioral control barriers in that an individual perceives flying as the only option open to whom.

The transportation tool factor will be depend to extent on the distance to the destination. This can also be interpreted in a social perspective as an intention with the resources available where much international tourism is structured around flying. To increase the availability of different transportation modes, tourists could choose holiday destination closer to home.

Finally, also how to predict future travel behavioral consumption. I feel that travel agents need to predict whether any country's random daily variation of weather factor is also important to influence travel behavior. e.g. in weather, temperature, rainfall and snowfall with traffic accidents factors will have relationship to cause travel demand.

Some scientists estimate suggest that when warmed temperatures and reduced snowfall are associated with a moderate decline in non-fatal accidents, they are also associated with a significant increase in fatal accidents. Thus increase in fatalities and temperature. Half of the estimated effect of temperature on fatalities is due to changes in the exposure to pedestrians, bicyclists and motorcyclists as temperature increase.

So, if any countries have rainfall, snowfall and low temperature to cause traffic accidents, whether this accident occurrence will influence the travelers who liking climb snow hills, riding bicycle, running sports who will avoid to travel to these countries‘ bad weather after occurs. So, why I feel that this natural climate factor will also be one serious factor to influence travel behavioral consumption. However, (AI) big data tool can predict more accurate than survey method when climate change to influence the country's climate to be poor, then it can predict when which countries are not popular acceptable to global country consumers' travel choice next month.

How can apply (AI) to provide travelling businesses with better-informed decisions

I shall explain how (AI) big data gathering technology can provide travelling businesses with better-informed decisions to drive top-line growth, deliver meaningful experience for travelling customers and smooth their path along the travelling consumer journey. The widely understood definition of (AI) involves the ability of machines or computers to learn human thinking, reasoning and decision-making abilities.

So, such as (AI) learning machine system can attempt to learn travelling consumer's travel destination or travel package thinking, judgement of their reasons why they choose to go to the destination to travel or why they choose to buy the travel package and learn how and why they make their past travelling decisions from their past travel big data gathering.

A Narrative science study in 2015 year identified that (AI) was being used primarily in voice recognition, machine learning virtual assistants and decision support. This study also highlighted the many branches of (AI) and that techniques and their definition are used interchangeably. It is possible that (AI) can be used to gather big data , then to analyze to help travel businesses to predict travelling consumer travel destination and travel package choice behaviors. For example, one of the most common techniques is traveler machine learning, where algorithms are used to perform tasks by learning from the airline or travel agent whose past all travelers‘ travelling destination choice and travel package choice historical data.

However, during 2017 year, search engines will begin to find what additional factors can influence past traveler personal travelling destination and travelling package travelling behavioral data into prediction of future travelling customer behavioral results, such as the online traveler (user's) history of travelling data searches, such as anywhere are the most popular travelling locations or travelling destinations and previously captures conservations.

Artificial intelligence will use this past travelling destinations and travelling package information to power predictive search results, e.g. predictive future travelling consumer's choice behavioral processing for where will be their preferable travelling destination choice and how to design travelling package to satisfy future travelling clients' needs.

Predictive search will improve the quality of online travelling search results, and provide new insights into travelling consumers‘ travelling destination and package behavior and the moments which matter to them. Search will give recommendation into tailored how travelling consumer individual travelling destination choice in travelling decision making process. Several of the largest online platforms already use (AI) travelling machine learning to improve predictive travelling consumer behavioral search results.

For example, Google's rank brain technology adds research by understanding the context in which the travelling consumer has entered it. Over time, rank brain will learn further from user behaviors Amazon's DSSTNE (pronouned

destiny) learns from shoppers' purchasing habits and consumption behavior to offer better product recommend actions, which Amazon can offer before a consumer has entered anything into the search bar.
Such as (AI) big data can gather past online travelers' e-ticket purchase transactions to conclude that online traveler's travelling choice habits and online traveler consumption behavior to offer better travelling destinations and travelling package opinions to travel agents or airlines. However, this technology is not independent of human input. For example, Google engineers will periodically retain the rank brain system to improve the models it uses.
For another example, in 2016 year , Apple computer revamped its travelling scene photos app to allow travelling consumers to search for specific travelling destinations in the travelling scene phots, they want to find anywhere travelling destination photos, not just dates and locations. Each travelling photo that an intelligent phone or intelligent pad user takes goes through 11 billion computations, so that travelling scene photos can understand exactly where is the travelling destination photography to let online travelling consumer to feel anywhere they plan to go to the location to travel. So, (AI) learning machine can make online travelling photos more attractive to influence potential travelers choose to the destination to travel after they see the travelling destination scene photos from internet.
It seems that in future, (AI) machine learning will allow online travelling search to evolve even further. Search engineers will deliver refined recommendations to airlines' online traveler e-ticket search users and use less human input to predict travelling consumers' needs from internet channel. For IBM computer example, it indicated 90% of the data that exists today has been created in the last two years.
This huge explosion of past traveler's e-ticket consumption data gives the opportunity to quickly spot and react to the latest trends, fashion and fads among its travelling clients and potential clients. This will allow airline or travel agent companies to better engage with younger travelling consumers, who gain influence access to the latest travelling destination and package trends.
They associate with to help define who they are as individuals. Thus, travelling company brands have to identify and make use of them before travelling consumers move on, but the vast quantity of past e-ticket purchase data available makes from internet channel. This a resource-intensive task. For next example, Lesara, a based online clothes store, uses this machine learning to inform its product decision often gathering information from internal and external sources.
When its trends -spotting shoes. Lesara has a range of over 20 styles and sells hundreds of pairs a day. It focus on giving consumers, the very latest trends allow Lesara to develop on average of 50,000 new items each year. It compared to 11,000 old items each year. Thus, travelling agents or airlines can attempt to apply (AI) big data gathering method to gather all past e-ticket purchase data, concerns where they prefer to choose to go to the destinations to travel and what travelling packages are the most attractive to the travelers to choose to buy. It aims to help them to predict where future travelers will prefer to choose to go to travel or what travelling package they will prefer to choose to buy next year.
For another (AI) big data prediction example, Lesara is one online clothes store, uses machine learning decisions after

gathering information from internal and external sources. One of its most popular products, shoes with LED started life when its trend spotting software flagged up a blogger wearing similar shoes. Now Lesara has a range of over 20 styles and sells hundreds of pairs a day. Its focus on giving consumers the very latest trends allows Lesara to develop an average of 50,000 new items each year, compared to 11,000 for its competitor Lara.

It seems (AI) big data gathering machine learning can help Lesara business to predict what kinds of shoes design or style that shoe consumers will prefer choose to buy in future shoe market trend. Thus, Lesara can predict shoe consumers' taste successfully and it can manufacture many attractive style of shoes.

(AI) machine learning can gather global past shoe consumer's shoe shopping experiences, then analyzes to make conclusion to give lesara recommendation successfully. This will make the experience more enjoyable for shoe consumers and allow Lesara to advert whose different new style or design of shoes to deliver them move relevant messages by understanding the context of the experience.

So, online travel agents or online airline can also attempt to apply (AI) big data gathering method to predict where travelers will prefer to go to travel and how they ought design travelling packages to attract them to choose to buy next year. Hence, (AI) big data gathering technology can conclude how to design traveler agents' travelling package products to be the most attractive to excite many travelers choose to buy their travelling package, due to it has more accurate to predict travelling consumer destination and travelling package choice behaviors to compare human themselves prediction judgement effort, e.g. travelling survey or marketing research, or telephone enquire. It seems that (AI) machine judgement effort is more accurate to compare to human judgment effort in travelling industry.

6.1 Future travel consumption behavior

Can (AI) big data gathering tool predict traveler individual habitual behavior , e.g. renting travel transportation tools ?

Can (AI) big data gathering tool can predict past traveler destination and travelling package choice habit and it can be intended to predict of future traveler behavior to people are creatures of habits judgement of future anywhere travelling destination choice next year or next month or next half year destination prediction ?

Many of human's everyday goal-directed behaviors are performed in a habitual fashion, the transportation made and route one takes to work, one's choice of breakfast. Habits are formed when using the some behavior frequently and a similar consistency in a similar context for the some purpose whether the individual past travel consumption model will be caused a habit to whom. e.g. choosing whom travel agent to buy air ticket or traveling package; choosing the same or similar countries' destinations to go to travel ; choosing the business class or normal (general) class of quality airlines to catch planes.

Does habitual rent traveling car tools use not lead to more resistance to change of travel mode? It has been argued that past behavior is the best predictor of future behavior to travel consumption. If individual traveler's past consumption behavior was always reasoned, then frequency of prior travel consumption behavior should only have an indirect link to the individual traveler's behavior. It seems that renting travel car tools to use is a habit example. So, a strong rent traveling car tools useful habit makes traveling mode choice. People with a strong renting of traveling car tools of habit should have low motivation to attend to gather any information about public transportation in their choice of travelling country for individual or family or friends members during their traveling journeys.

Even when persuasive communication changes the traveler whose attitudes and intention, in the case of individual traveler or family travelers with a strong renting travel car tools habit. It is difficult to change whose travel behaviors to choose to catch public transportation in whose any trips in any countries. However, understanding of travel behavior and the reasons for choosing one mode of transportation over another. The arguments for rent traveling car tools to use, including convenience, speed, comfort and individual freedom and well known.

Increasingly, psychological factors include such as, perceptions, identity, social norms and habit are being used to understand travel mode choice. Whether how many travel consumers will choose to rent traveling car tools during their trips in any countries. It is difficult to estimate the numbers. As the average level of renting travel car tools of dependence or attitudes to certain travel package policies from travel agents. Instead different people must be treated in different ways because who are motivated in different ways and who are motivated by different travel package policies ways from travel agents.

In conclusion, the factors influence whose traveler's individual traveler destination choice behavior The factors include either who chooses to rent traveling car tools or who chooses to catch public transportation when who individual goes to travel in alone trip or family trip. It include influence mode choice factors, such as social psychology factor and marketing on segmentation factor both to influence whose transportation choice of behavior in whose trip. So, (AI) big data can be attempted to gather past traveler transportation tool choice, rent travelling car tools choice or catching public transportation tools choice to predict where destination can provide what kind of transportation tool to attract many travelers to choose to go to the place to travel.

How (AI) big data determine future travel behavior from past travel experience and perceptions of risk and safety for the benefits to travel consumers?

How (AI) big data determine future travel behavior from past travel experience and perceptions of risk and safety for the benefits to travel consumers? Why does individual traveler avoid certain destination(s) is(are) as relevant to tourist decision making as why who chooses to travel to others?

Perceptions of risk and safety and travel experience are likely to influence travel decisions. If travel agents had efforts to predict future travel behavior to guess whether travelers will feel where is(are) risk and unsafe to cause who does not choose to go to the country to travel. Then, the travel agents will avoid to choose to spend much time to design the different traveling package to attract their potential travel consumers to choose to travel. The reason is because in the case of individual traveler's tourism experience, the traveler whose past disappointment travel experience (psychological risk) will be a serious threat to the traveler's health or life (health, physical or terrorism risk). The past safety or unhealthy risk to the country(countries) will influence the traveler decides to choose not to go to the countries(country) to travel again in the future.

6.2 What is push and pull factors to influence any traveler who chooses where is whose preferable travelling destination ?

How to apply (AI) big data to predict individual traveler's behavioral intention of choosing a travel destination? Understanding why people travel and what factors influence their behavioral intention of choosing a travel destination is beneficial to tourism planning and marketing. In general, an individual's choice of a travel destination into two forces.

The first force is the push factor that pushes an individual away from home and attempt to develop a general desire to go somewhere, without specifying where that may be.

The other force is the pull factor that pull an individual toward in destination, due to a region-specific or perceived attractiveness of a destination. The respective push and pull factors illustrate that people travel because who are pushed by whose internal motives and pulled by external forced of a destination. However, the decision making process leading to the choice of a travel destination is a very complex process.

For example, a Taiwanese traveler who might either choose new travel destination of Hong Kong or another old travel Asia destinations again or who also might choose any one of Western country, as a new travel destination. The travel agents can predict where who will have intention to choose to travel from whose past behavior and attitude, subjective and perceived behavioral control model. When (AI) big data gather past every country traveler number who chose to go to which countries to travel in order to judge where destinations will be the country travelers' travelling choice destinations in the future.

The factors influence where is the traveler choice, include personal safety, scenic beauty, cultural interest, climate changing, transportation tools, friendliness of local people, price of trip, trip package service in hotels and restaurants, quality and variety of food and shopping facilities and services etc. needs. So, whose factors will influence where is the individual travel's choice. It seems every traveler whose choice of travel process, will include past behavior. e.g. travelling experience, travelling habit, then to choose the best seasoned travelling action to satisfy whose travel needs. This process is the individual traveler's psychological choice process, who must need time to gather information to compare concerning of different travel packages, destination scene, climate change, transportation tools available to the destination, air ticket price etc. these factors, then to judge where is the best right destination to travel in the right time.

Hence, (AI) big data can gather past different countries' climate changing data, transportation tool changing data, destination scene environment changing etc. different data to give opinions to travelling businesses whether any country's these above factors will influence about how many traveler number will be increase or decrease in the future.

6.3 Why can expectation, motivation and attitude factor influence travelling behavior?

Social psychology is concerned with gaining insight into the psychological of socially relevant behaviors and the processes. For instance, on a global level bad influence to global warming, it influences some countries extreme cold or hot bad climate changing occurrence, then it ought influence some travelers' behavioral decision to change their mind to choose some countries to go to travel at the moment which do not occur extreme hot or cold climate (temperature). e.g. above than 40 degree in summer or below than 0 degree in winter. Due to the extreme climate changing environment in the countries, it will cause them to feel uncomfortable to play during their trips. So, the global warming causes to climate changing factor will influence the numbers of travel consumption to be reduced possibly. This is global climate changing environment factor influences to bad or uncomfortable social psychological feeling to global travelers' mind of traveling decision. What is individual traveler expectation, motivation and attitude? Tourism sector includes inbound (domestic) tourism and outbound (overseas) tourism both incomes to any countries. According to recent article, a tourist behavior model has been developed, called the expectation, motivation and attitude (EMA) model (Hsu et al., 2010).

This model focuses on the pre-visit stage of tourists by modeling the behavioral process by incorporating expectation, motivation and attitude. Travel motivation is considered as an essential component of the behavioral process, which has been increasing attention from the travel; industry. The economic approach defines "tourism" is an identifiable nationally important industry. It includes the component activities of transportation, accommodation, recreation, food and related service. So, tourism behavioral consumption is concerned the individual tourist's usual habituate of the industry which responds to whose needs, and of the impacts that both the tourist and the tourism industry have on the socio-cultural, economic and physical environment.

However, travel motivation means how to understand and predict factors that influence travel decision making. According to Backman and others (1995, p.15), motivation is conceptually viewed as " a state of need, a condition that services as a driving force to display different kind of behavior toward certain types of activities, developing preferences, arriving at some expected satisfactory outcome." So, motivation and expectancy which has close relationship to any tourist before who decided to do any tourism of behavior.

Some economists confirmed motivation and expectancy which has relations, such as expectation of visiting an outbound destination has a direct effect on motivation to visit the destination; motivation has a direct effect on attitude toward visiting the destination; expectation of visiting the outbound destination has a direct affection on attitude toward visiting the destination and motivation has a mediating effect on the relationship in between expectation and attitude.

Hence, (AI) big data can gather all the country's climate environment change, transportation tool change, entertainment scene change, hotel price and restaurant price change etc. data to give opinions whether the country will attract how many traveler to choose to go to travel in the year.

What is (AI) deep learning techniques to forecast travelling environment behavioral consumption

Prediction how many travelers will choose to go to the country to travel. It is similar to apply deep-learning technology to predict how to raise the agricultural farming productivity in the agricultural export country.

The (AI) deep-learning technology leads to performance enhancement and generalization of artificial intelligent

technology. It influences the global leader in the field of information technology has declared its intention to utilize the deep-learning technology to solve environmental problems, such as climate change.
So, it will help agriculture farming businesses can raise any plant food: vegetable, fruit, rice which grow up very easily if farmers can apply (AI) deep-learning technology to solve environment problems to influence their plant food grow. If the whole year seasonal change is very good and it is suitable for any plant food to grow in farming land easily, e.g. rain is enough and soil is enough for any plant food to grow in the farm lands. Then, fruit, rice, vegetable etc. agriculture businesses will have much beneficial attribution to global farmers.
The question is how to use deep-learning technologies in the environmental field to predict the status of pro-environmental consumption. We predicted the pro-environmental consumption index based on Google search query data, using a recurrent neural network (RNN model). To certify the accuracy of the index, we compared the prediction accuracy of the RNN model with that of the ordinary least square and artificial necessary network models.

For example, the RNN model predicts the pro-environmental consumption index better than any other model. we expect the RNN model to perform still better in a big data environment because the deep-learning technologies would be increasingly as the volume of data grows. So, deep-learning technologies could be useful in environmental forecasting to prevent damage caused by climate change to influence any rice, vegetable, tomato, potato, fruit etc. different plant food grow in any countries' farming land easily.
For South Korea example, over 800 government agencies spent 2.2 trillion Korea won on eco-products in 2014 year. However, green products are rarely purchased outside these agencies. This phenomenon occurs because there is a gap between consumer attitudes and behavior , that is environmental attitude is a major factor in decision making vis-a-vis the consumption of " green" food and services (Jorea Ministry of Environment, 2015).
Therefore, it is necessary to understand those consumer attitude, that will lead to sustainability-conductive behavior and consumption. (AI) Deep learning system can be applied to attempt understand those traveler attitude to environment protection to fly to which country. For example, (AI) deep learning system can attempt to gather data concerns how many Hong Kong people concern air pollution challenge to influence their health, then it can attempt to predict how many Hong Kong travelers do not choose to go China travel, due to the air pollution challenge to influence their health.

7.1 Environmental travel consumption prediction

Recently, many researchers have studied pro-environmental consumption and household indexes as well as suicide rate predictions using messages posted by internet users on Google trend, Tweets etc. channel.
Whether can environmental consumption be predicted by (AI) deep-learning technological internet channel to influence how many travelers choose to go to the country to travel?
How can impact the pro-environmental consumption attitudes of green policies to influence how many travelers choose to go to the country to travel?

For example, Korea scientists estimated pro-environmental attitudes using search query data provided by Google trend and confirmed through regression analysis, that pro-environmental attitude has a positive correlation with the pro-environmental attitude index. They also explained that environment-friendly attitude of residents plan an important role in policy making. In the past, most household consumption indexed were calculated through surveys, but (AI) deep-learning technological tool " big data" have recently gained research attention (Lee et al. 2016). So, (AI) deep learning technology can attempt to gather whether how many Korea residents who concern environment pollution to influence their eating green food attitude then to judge whether how many Korea residents hope to leave their country to travel anywhere either high risk environment pollution countries to travel or low risk environment pollution countries to travel in the future.

It seems that (AI) deep-learning technology can help agricultural export countries' farmers , e.g. US, UK, Canada, New Zealand, Australia, Japan, China, India etc. they can predict environmental behavioral consumption to any rice, tomato, potato , fruit, vegetable etc. plant food consumers. The beneficial advantages to them include as below:

(a) Assuming they know their countries' weather, when it has less rain to cause drought or when it has more rain in any seasonal time in the year. They can choose not to grow any kinds of above these plant food to avoid loss.

(b) They can make any kinds of above these plant food price raising after their prediction of these bad seasonal time to cause their plant food shortage supply challenge. Because these plant food consumers' demand number is more, but the supply of these above plant food supply number is less. However, due to they had predicted when the bad seasonal time can not allow them to grow these above plant food before. So, they have enough time to grow many these above plant food number in predictive good seasonal time to prepare to supply to their plant food import countries' plant food consumers to eat. Thus, these predictive environmental consumption plant food export countries can raise their plant food price to sell to them. When, the other non-pre-predictive environmental consumption plant food export countries can not supply any one of those plant food to them to eat, due to the bad climate to cause them can't grow any one of these plant food to export to sell.

Thus, (AI) deep-learning technology can be applied to predict how to raise the plant food supply number in order to raise price to the import plant food countries consumers to eat, due to they feel difficult to buy these plant food to eat in the bad climate seasonal time in whole year.

(c) (AI) deep-learning technology can help climate scientists to find what reasons cause their countries; rain sudden increases or cause their countries' rain sudden decreases. After its gathering data analysis, it can assist climate scientists to find solution methods to attempt to control the rain level can be right falling down level to let agricultural export farmers who can grow their plant food to sell to agricultural import countries in whole year.

(d) The agricultural export countries' farmers can apply (AI) deep-learning technology to help them to choose whether growing which kinds of plant food in that whether climate time to earn more plant food consumption number more easily.

Due to the agricultural countries climate will often change, for example, tomato, potato, rice, fruit etc. plant food can be adapt to grow in more rain time, but vegetable can not be adapt to grow in more rain time. If farmers can apply this technology to predict when it will have move rain or when it will have less rain to fall down in their countries. Then, they can choose to grow which kinds of plant food number more, in the suitable seasonal climate time in order to raise plant food growing number productivities to supply to sell to satisfy any agricultural food import countries' demand effectively.

(e) (AI) deep-learning technology can help agricultural import countries to solve agricultural food shortage challenge in long term. When this technology can be popular to base applied by the agricultural plant food export countries. It will solve global agricultural food shortage challenge. For example, when one agricultural export countries' farmers can popular accept to apply this technology to predict when to grow which kinds of plant food more to rise number productivities to sell. e.g. vegetable, fruit, rice Besides another agricultural export countries' farmers can also accept to apply this technology to predict when to grow plant food, e.g. potato, tomato to raise number productivities to sell. Then, they can concentrate on growing the specific kinds of plant food in order to raise the specific plant food number productivities in every seasonal change time every month. Then, global agricultural plant food supply must be raised, due to these predictive environmental change farmers can know who ought grow which kinds of plant food to sell to raise number productivities.

Consequently, (AI) deep learning can gather where countries will have high risk environment pollution to influence health food supply. Then, it can give opinions to travelling businesses when these high risk environment pollution countries will encounter the traveler number to be decreased, due to the environment pollution serious challenge will occur.

7.2 What methods can predict future travel behavioral consumption ?

How to use qualitative of travel behavioral method to predict future travel consumption from (AI) big data ?

I also suggest to use qualitative of travel behavioral method to predict future travel consumption. Methods such as focus groups interviews and participant observer techniques can be used with quantitative approaches on their own to fill the gaps left by quantitative techniques. These insights have contributed to the development of increasingly sophisticated models to forecast travel behavior and predict changes in behavior in response to change in the transportation system. I shall indicate the weaknesses of human travelling investigation methods as below:

First, survey methods restrict not only the question frame but the answer frame as well, anticipating the important issues and questions and the responses. However, these surveys methods are not well suited to exploratory areas of research where issues remain unidentified and the researched seek to answer the question "why?".

Second, data collection methods using traditional travel diaries or telephone recruitment can under represent certain segments of the population, particularly the older persons with little education, minorities and the poor. Before the survey, focus group for example can be used to identify what socio-demographic variables to include in the survey, how best to structure the diary, even what incentives will be most effective in increasing the response rate.

After the survey, focus, focus groups can be used to build explanations for the survey results to identify the "why" of the results as well as the implications. One Asia Pacific survey research result was made by tourism market investigation before. It indicated the travel in Asia Pacific market in the past, had often been undertaken in large groups through leisure package sold in bulk, or in large organized business groups, future travelers will be in smaller groups or alone, and for a much wider range of reasons.

Significant new traveler segments, such as female business traveler. The small business traveler and the senior traveler, all of which have different aspirations and requirements from the travel experience.

Moreover, Asia tourism market will start to exist behaviors in the adoption of newer technologies, a giving the traveler new ways to manage the travel experience, creating new behaviors. This with provide new opportunities for travel providers. The use of mobile devices, smartphones, tablets etc. and social media are the obvious findings to become an integral part of the travel experience. Thus, quality method can attempt to predict Asia Pacific tourism market development in the future. It is such as (AI) big data gathering tool can give traveler quality opinions to any travelling businesses to make the more accurate where will be the popular travel destination choice next month or next half year or next year.

However, improving the predictive power of travel behavior models and to increase understanding travel behavior which lies in the use of panel data(repeated measures from the same individuals). Whereas, cross-sectional data only reveal inter-individual differences at one moment in time, panel data can reveal intra-individual changes over time. In effect, panel data are generally better suited to understand and predict (changes in) travel behavior. However, a substantial proportion was also observed to transition between very different activity/travel patterns over time, indicating that from one year to the next, many people renegotiated their activity/travel patterns.

7.3 How to apply advanced traveler information systems (ATIS) to predict future travelling behavior?

Nowadays, information can impact on traveler behavior and network performance. For example, when steadily growing levels of vehicle ownership and vehicle miles traveled information has been identified as a potential strategy towards man aging travel demand, optimizing transportation networks and better utilizing available capacity. Toward, this goal to predict further tourist behavioral consumption. Many countries, government tourism development institutes has applied advanced traveler information systems (ATIS) which travel behavior models and high-fidelity network performance models made increasingly feasible through the rapid advances in computer power. Crucial components of this problem domain are the modeling of individual tourist drivers' response to travel information and the development accurate guidance of relevance to real would trip makers. So, this advanced traveler information systems (ATIS) can assist the tourist who like to rent travelling car tools to travel in any countries own free traveler information systems service conveniently. Also, this travel information system can be intended to assist travelers to make better travel choices. e.g. this system can improve the decision making of individual traveler rather than improvements of network performance overall. So, we need to understand how tourists make their travel plans. Also, understanding decision process that lead to booking of the trip is equally important, as it allows of a potential behavior.

7.4 How can online tourism sale channel influence traveling consumption of behavior?

Nowadays, internet is popular, it seems that booking air ticket behavior of using internet is predicted to influence overall tourism air tickets payment method. Tourism industry has grown in the previous several decades. Despite its global impact, questions related to better understanding of tourists and whose habits. Using online travel air ticket booking benefits include booking electronic air tickets can be made from entering any electronic travel agents websites in the short time and electronic travel ticket payers do not need leave home, who can pay visa card to pre booking any electronic travel ticket from online channel conveniently.

7.5 How can analyze activity based travel demand ?

Nowadays, human are concerning the traffic congestion and air quality deterioration, the supply oriented focus of transportation planning has expanded to include how to manage travel demand within the available transportation supply. Consequently, there has been an increasing interest in travel demand management strategies, such as congestion pricing that attempts to change aggregate travel demand. The prediction aggregate level, long term travel demand to understanding disaggregate level (i.e. individual levels) behavioral responses to short term demand policies, such as ride sharing incentives, congestion pricing and employer based demand management schemes, alternate work schedules, telecommuting limitation of travel agent traditionally work nature shall influence oriented trip based travel modelling passenger travel demand indirectly.

Finally, online travel purchase will be popular to influence the number of travel behavioral consumption nowadays. Any travel package products can be sold from websites to attract travelers to choose to pre-book air ticket for any trips conveniently. In the past ten years, the internet has become the predominant carrier of all types of information and transactions. Regarding travel decisions, internet has also become an important sales channels for the travel industry, because it is associated with comparably lower distribution and sales costs, but also because it adapts to high supply and demand dynamics in this industry. Consequently, the travel and tourism industry tries to increase the internet sale specific share of sales volumes. So, internet sale channel has changed travel consumption behavioral pattern and characteristics and travel experience. For example, Switzerland has one of the highest population-to-computer ratio in Europe. It is also one of the most highly internet penetrated countries in terms of use of the WWW on a day-to-day basis, with more than 75 percent of the population older than 14 years using the WWW daily (ICT, 2005).

The reason of booking online tourism may include: convenience, fast transaction, finding traveling package choice easily, more airline seats available. So, online booking tourism will influence the traditional tourism agents visiting of sales and air tickets and travelling package numbers to be decreased. Finally, the online booking tourism market shares will be expanded to more than traditional tourism agents visits sale market in the future one day. So, the travel agents who still use the traditional tourism visiting sale channel which ought raise whose features to compare to differ to online tourism sale channel if these traditional tourism agents want to keep competitive ability in tourism industry for long term.

7.6 What is actively based patterns of urban population of travel behavioral prediction method?

Actively based patterns of urban population. It is a method of motivational framework means in which societal constraints and inherent individual motivations interact to shape activity participation patterns. It can be used to predict one city or urban the numbers of travel demand in the year. It has two elements: First, capability constraints refer to constraints are imposed by biological needs, such as eating and sleeping and/or resources, such as income, availability of cars etc. to undertake the urban or city's family activities in the year. Second, coupling constraints define where, when and the duration of planning activities that are to be pursued with other individuals. So, this method needs to gather information (data) to get the relationship between activities, travel and spending work time and space time to evaluate whether there are how many families who have real needs to spend time to go to travel in the year.

7.7 What is trip based versus activity based approaches?

What is trip based versus activity based approaches? The fundamental difference between the trip-based and activity based approaches is that the former approach directly focuses on trips without explicit recognition of the motivation or reason for the trips and travel. The activity based approach , on the other hand, views travel as a demand derived from the need to pursue travel activities. So, it is better understand the individual or family behavior basis for individual or family travelling decision regarding participation in travelling activities in certain places or cities or countries at given times and hence the resulting travel needs. This behavioral basis includes all the factors that influence the why, how, when and where of performed activities and resulting individuals and household, the cultural/social norms of the community and the travel surrounding environment.

Another difference between the two approaches is in the way travel is represented. The trip based approach represents travel as a collection of trips. Each trip is considered as independent of other trips, without considering the inter-relationship in the choice attributes , such as time, destination and mode of different trips. As tours are chains of trips beginning and ending at a same location , say home or work. The tour based representation helps maintain the consistency across and capture the interdependency and consistency of the modeled choice attributed among the trips of the same tour.

In addition to the tour based representation of travel, the activity based approach focuses on sequences or patterns of activity participation and travel behavior, using the whole day or longer periods of time is the unit of analysis. Such as approach can address travel demand management issues through an examination of how people modify their activity participation, for example, will individuals substitute more out-of-home activities for in home activities in the evening of who arrived early form work due-to a work schedule change?

The major difference between trip based and the activity based approaches is in the way, the time dimension of activities and travel is considered. In the trip based approach, time is reduced to being simply a cost making a trip and a day's viewed as a combination, defined peak and off peak time periods. On the other hand, activity based approach views individuals' activity travel patterns are a result of their time use decisions with a continuous time domain. As individuals have 24 hours in a day or multiples of 24 hours for longer periods of time and decide how to use that travel among or allocate that time to activities and travel and with who, subject to their socio-demographic, transportation system and other and scheduling of trips. So, determining the impact of travel demand management policies on time use behavior is an important step to assessing the impact of such policies on individual travel behavior. The final major difference between this two approaches relates to the level of aggregation. In the trip based approach, most aspect of travel, e.g. number of trips etc. are analyzed at an aggregate level.

Consequently, trip based methods accommodate the effect of socio-demographic attributes of households and individuals in a very limited fashion, which limits the activity of the method to evaluate travel impacts of long term socio-demographic characteristics of the individuals who actually make the activity travel choices and the travel service characteristics of the surrounding environment. So, the activity based models are better equipped to forecast the longer term changes in travel demand in response composition and the travel environment of urban areas. Also, using activity based models, the impact of policies can be assessed by predicting individual level behavioral responses instead of employing trip based statistical averages that are aggregated over defined demographic segments.

7.8 Can apply (AI) big data gathering method predict senior age will be main travelling target?

In the past, Germany government had established tourism survey analysis to analyze survey data in order to arrive at reliable conclusions on future trends in travel behavior. To aim to find how demographic change will influence the tourism market and how the industry can adapt to those changes. The travel analysis provided data on tourism consumer behavior, including attitudes, motives and intentions. Since, 1970 year, it is based on a random sample, representative for the population in private households aged 14 years or older. Then, a continuous high scientific standard combined with a national and international users makes the travel analysis a useful tool and reliable source for tourism industry and policy decisions. It aimed to gather statistical data. e.g. on the age structure and on demographic trends, quantitative and qualitative analysis with time series data from the travel analysis. It shows e.g. not only the future volume , quite different from today's seniors, or how who will travel of family holidays will change, e.g. single parents of low, but grandparents of growing significance for tourism.

Demographic change is said to be one of the important drivers for new trends in consumer traveling change behavior in most European countries (e.g. Lind 2001). Because the growing number of senior citizens in the European Union and other industrialized countries, such as the USA and Japan, looks to become one of the major marketing challenges for the tourism industry. United Nations statistics predict that the share of people being 60 age or older will grow dramatically in the coming future, and is expected to rise from 10 percent of the world population in 2000 year to more than 20 percent in 2050 year (United Nations Population Division, 2001). From its statistic, some data showed that travel propensity increased throughout life until the age of about 50 years of age and was then kept stable until very late in life 75 age. The most important results is that the travel propensity when getting older is not going down between 65 and 75 age of course, the overall development of this variable is influenced by a lot of other factors which are responsible for quite a variation over time. It is now possible to suggest that the general pattern of travel propensity is one of the key indicators for holiday life cycle travel behavior, includes three stages. The growth stage tends to increase from early adult hood until 45 age old or when reaching some 80%. The next stage is stabilization from the ages of around 50 age, until 75 age old, starting with a lower increase. Finally, the decrease stage is a slight decrease occurs once people reach the more advanced age of 75 age to 85 age old (Lohmann & Danielsson 2001).

So, it seems Germany government tourism prediction to future travelers' behavior indicated these findings, such as on how future senior generations will travel, who had used survey data to examine the patterns of travel behavior of a generation getting older and applied the findings to draw conclusions on the future. Also, it predicted that on the future of family trips, family segmentation will be the travel behavior patterns in the future. These findings together with the statistical data on demographic change allowed for a better understanding of the coming tends in family holidays. It's aim developed in consumer behavior related to demographic change and predicted what will happen future of tourism one had to consider other influences and drivers as well, for example, trends on the supply side. e.g. low cost airlines or in travelling consumption behavior in general whether how the past may provide a key to predict travel patterns of senior citizens to the future.

Given the projected growth of the senior citizens market, designing specific marketing strategies to meet the prospective needs of elderly tourists will become increasingly important. It has been an implicit assumption that it will be a close relationship between the travel behavior of today's senior citizens and the those of future ones. The growing number of senior citizens in the world. e.g. China, Hong Kong, Japan, USA etc. countries. Global senior citizen tourism market will be based solely on demographic predictions about the future of the population's age structure. However, many of these seniors won't only live longer but will be fitter and more active until later in life. Many of the will also have plenty in life. Many of them will also have plenty of time and money to spend on travel. So, will these new seniors behave like today's senior citizens? Will they adopt the same travel behavior as the previous generation or become a new market of oldies for the leisure and tourism industry? However, to determine the actual number of senior citizens who will be travelling and to sought to evaluate and specify certain difficult to predict the actual numbers of senior citizen to any country. However, they can be based on the implicit assumption that there is a close relationship between the travel behavior of past, present and future seniors. But is this a valid assumption? As the revise- analysis travel analysis survey, which was conducted in Germany every year, offered some interesting data possibilities. It was designed to monitor the holiday travel behavior, opinions and attitudes of Germans and has been carried out since 1970 year, questions in the questionnaire. Data are based on face to face interviews, with a representative sample of more than 7,500 respondents, the interviews being carried out in January each year. All results refer to the average for the defined generated, which ranges generally over ten years. The group of people then at the age of 60 to 69 age is described. This corresponds to the same generation ten years ago, when they had an age of 50 to 59 age. When this methodological approach is not necessarily very sophisticated, it does have the important advantages of being cost effective.

7.9 IS (AI) big data gathering method a better psychological method to compare human marketing research method predict travel behavioral consumption?

On the psychological view point, I think individual traveler's character will have those kind of personal characteristics. First, simplicity searchers value above everything ease not transparency in their travel planning and holiday making, and are willing to avoid having to go through extensive research. Second, cultural purists use their travel as an opportunity to immerse themselves in an unfamiliar looking to break themselves entirely from their home lives and engage. Sincerely with a different way of living. Third, social capital seekers understand that to be well travelled is a personal quality, and their choices are shaped by their desire to take maximum of social reward from their travel. They will exploit the potential of digital media to enrich and inform their experiences, and structure their adventures always keeping in mind they are being watched by online audiences. Finally, reward hunters seek a return on the investment who make in their busy , high-achieving lives. Linked in part to the growing trend of wellness, including both physical and mental self-improvement who seek truly extraordinary and often indulgent or luxurious' must have experiences.

Why needs to know the personal character of individual traveler's characteristics? Because if travel agents could feel which kinds of individual traveler's character, then who can predict which kind of travel package to design to them more easily. For example, how to determine future travel behavior from past travel experience and perceptions of risk and safety? We need to concern that the influences of past international travel experience, types of risk associated with international travel and the overall degree of safety feeling during international travel on individual's travelling experiences likelihood of travelling to various geographic regions on their next international vacation trip or avoidance of those regions, due to perceived risk. Because individual traveler's experience of safety risk degree to the countries, it will influence who chooses to go to the countries/country to travel again.

Why travelers avoid certain destinations are as relevant decision making as why who choose to go to the country(countries) to travel. Perceptions of risk and safety and travel experiences are likely to influence travel decisions; efforts to predict future travel behavior can benefit to individual tourist's decision making.

As Weber & Bottorn (1989) defined risky decision is as "choices among alternatives that can be described by probability distributions over possible outcomes" (p.114). Some psychologists judge subjective perceptions of physical reality, i.e. image of a particular tourist destination, whereas value judgement refers to the way individual rank destinations according to whose attributes. i.e. attractiveness, safety, risk etc. factors to form on overall image. So, if the individual traveler had unhappy and worried and unsafe experiences to go to where the place(country) to travel during whose vacation time before. Then, this negative travel experience will influence who is afraid to go to the place (country) to travel again. Risk of place, country, destination or region means the danger is relatively high to the place, i.e. increasing in airplane accidents, crime or terrorist activity targeting citizens of potential traveler's nationality or the probability of occurrence is great , i.e. recent occurrences involving travel regions/destinations under consideration or effective actions to control consequences exist. i.e. selecting safe regions and destinations, taking extra precautions when traveling to risky destinations. These risk factors will influence the individual traveler who chooses to cancel travel plan to go to the country again.

Another interesting research, how to predict behavioral intention of choosing a travel destination, which has focus of tourism research for years, but the complex decision making process leading to the choice of a travel destination has not been well researched. The planned behavior model using its core constructs, attitude, subjective norm and perceived behavioral control, with the addition of the past behavioral variable on behavioral intention of choosing a travel destination.

Understanding why people travel and what factors influence their behavioral intention of choosing a travel destination is beneficial to tourism planning and marketing. Understanding travel motivation is the push and pull model. The idea of the push and pull model is the decomposition of an individual's choice of a travel destination into two forces. The first force is the push factor that pushes an individual away home and attempts to develop a general desire to go somewhere else, without specifying where that may be. The second force is the pull factor, that pulls on individual toward a destination, due to a region specific travel location or perceived attractiveness of a destination. The respective push and pull factors illustrate that people travel because who are pushed by their internal motives and pulled by external forces of a destination. Nevertheless, how push and pull factors guide people's attitude and how these attributes lead to behavioral intentions of choosing a travel destination have rarely been investigated. The decision making process leading to the choice of a travel destination is a very complex process. The planned behavior model is as a research framework to predict the behavioral intention of choosing a travel destination. The model based on the three constructs of attitude, subjective norm, and perceived behavioral control (Fishbein & Ajzen, 1975).

In conclusion, the factors can influence travelers who decide to choose to travel the country, which include personal safety was perceived to the highest motivation factors among the important factors which include, scenic beauty, cultural interests, friendliness of local people, price of trip, services in hotels and restaurants, quality and variety of food and shopping facilities and services. The factors include both push and pull. Push factors include knowledge, prestige, and enhancement of human relationship etc., whereas, the most significant pull factors include high technologic image, expenditure and accessibility etc. For example, Japanese travelers visiting Hong Kong. Push factors are such as exploration dream fulfillment and pull factors are such as benefits sought, attractions and good climate city. It will be the factor of future travel patterns and motivations of sub-cultural and ethic groups for Japanese choice to go to Hong Kong travelling.

7.10 How can apply (AI) digital channel (big data gathering method) predict travelling consumer behaviors?

(AI) big data digital channel can be applied to help travelling businesses to evaluate whether how much the e-ticket price and travelling package price is the most attractive or reasonable to persuade travelling consumers feel it is the most reasonable price to choose to buy the airline's e-tickets or the travel agent's travelling package product from internet channel . It helps travelling consumers to feel which airlines or travelling agents which ought change their e-ticket and/or travelling package price to let travelling consumers to choose to buy the airline e-ticket or the travelling agent's travelling package products from internet channel. It can be applied to predict whether how many travelling consumer numbers can be increased or decreased when the airline e-ticket price is variable or the travelling agent travelling package price is variable . It aims to give opinions to help any online airlines or travelling agents to judge whether which e-ticket or travelling package price is the most reasonable to let travelling consumers to accept to choose to buy which airline's e-tickets or traveling agent's package products more attractive.

Thus, (AI) e-ticket or e-travelling package price measurement technology can be preference to be applied online communication ecommerce and mobile phone internet platform aspect. As traveling businesses can enter their past e-ticket or travelling package prices data and past travelling customer number data into computer or mobile. Then, (AI) price measurement technology can gather these data to analyze these e-ticket or travelling package product prices and past travelling customer number to compare their e-ticket and/or travelling package prices variable changing range level to find their e-ticket and /or travelling package price variable difference to measure to make conclusion about every travelling package or/and e-ticket product's price variable changing will influence how many travelling customer number increase or decrease changing to choose to sell their different kinds of travelling package or e-ticket products more accurate. Then, (AI) price measurement software will help them to analyze all past e-ticket and/or travelling package price variable changing data to compare whether which e-ticket and/or travelling package price range can let travelling customers to feel it is more reasonable and attractive to influence them to choose to buy their e-ticket or travelling package product among different airlines and travel agent choices. Because any e-ticket or travelling package product's price is one important factor to influence travelling consumers to choose to buy the airline's e-tickets or travelling agent's travelling package products.

For example, Amazon publish has applied (AI) price measurement technology to help authors to decide how much every different topic of e-book or paper book price, it can attract the largest number of readers to buy. Any one author only needs to type whose book name to Amazon publish author himself/herself Amazon website. Amazon publish (AI) price measurement learning machine will help them to auto-calculate and judge how much e-book or paper book price is the most attractive and the most reasonable in order to increase reader number to buy their e-books or paper books to read. So, (AI) online price measurement machine will gather past similar book names and past every similar book readers' reading times and the number of readers to give opinions to let every author to judge whether his/her very new e-book or paper book ought charge how much price to the e-book or paper book which can attract many readers to choose to buy. Although, it is not ensure that the e-book or paper book price must let readers to feel it is the most reasonable price to choose to buy in reader's view point. However, it has other factors to influence readers' choice to buy the e-book or paper book, e.g. whether the book content is attractive to public, the author's familiarity, the book's page is enough or not to satisfy readers to read etc. factors. But, instead of all these extra factors to influence readers to choose to buy the book to read. (AI) price measurement learning machine can real give opinions to every author to let them to judge the e-book or paper book different price range whether is too high to influence readers to choose to buy to read or tool low to influence readers feel it is possible poor content book to compare other similar content books. Thus, (AI) price measurement machine can help authors to predict every reader's reading behaviors or reading experience and reading habit from online channel in short time easily. The author only enter the book name to let Amazon publish price measurement machine to check, it will follow past reader's reading habit and reading experience to judge whether the similar all book topic sale record to judge how much price is the reasonable price to attract many readers to buy the book.

Hence, (AI) can be applied to digital channel to help travelling businesses to predict travelling consumer behavior in the future. In the future, mobile/smartphone, laptop, desktop will be most frequent used ecommerce channels to develop online business. So, (AI) can be also applied to these platforms to gather data to make analysis to help travelling businesses to predict travelling consumer purchase behaviors popularly. Due to , ecommerce is popular to global, so digital online and instore channels can be one good channel to let (AI) learning machine to make platform to gather past every online travelling consumer purchase (buying) experience data to help travelling businesses to build airline or travelling agent brand personality and having a responsible, positive impact on society.

To apply (AI) learning machine technology to understand travelling customer online purchase behavior, it will raise business e-commerce successful chance: For example, (AI) learning machine can help travelling businesses to gather data to analyze to determine whether short-term or long-term signals in the online travelling consumer behavior that indicate higher purchase intents to let every online travelling business to know. (AI) learning machine can find that online users with long-term purchasing intent tend to save and click through on more content.

However, as online travelling users approach the time of purchase their activity becomes more topically focused and actions shift from saves to searches from online travelling consumption channel. Then, (AI) learning machine will further find that the brand airline and/or travelling agent purchase signals in online travelling consumption behavior can exist weakness before an online travelling purchase is made and can also be traced across different online travelling purchase categories. Finally, (AI) learning machine synthesize these insights in predictive models of online travelling user purchasing intent to the brand of airline or/and travelling agent travelling package product. Taken together, it's work identifies a set of general principles and signals that can be used to model online travelling user e-ticket and/or travelling package purchasing intent across many online content discovery applications. Thus, (AI) learning machine can help online travelling businesses to gather any online travelling users' click online travelling behaviors data to judge whether there are how many online travelling users will choose to find their online travelling business websites to make final decisions to buy their travelling package or/and e-ticket products from online channels. Then, it will give opinions to help the online travelling businesses to let it to judge whether what are the important website factors will help its online travelling business to attract many online travelling consumers, e.g. designing unattractive travelling website issue, online unattractive scene photos issue, unclear website travelling photo color issue, unclear website travelling advertisement message, contents and words impressions issue, lacking image movement frequent attractive seeing issue etc. different website factors. Thus, online digital channel will be one good choice to apply (AI) learning machine to help travelling businesses to predict travelling consumer behaviors.

Thus, (AI) big data technology can also assist travelling consumers to gather different manufacturers' data to compare what their advantages and disadvantages of their travelling package products are. Then, travelling consumers can make comparison to choose which airline or travelling agent is the suitable to whom to buy e-ticket or pre-booking travelling package in online travelling consumption market.

.

Thus, I believe that artificial intelligent "big data" gathering method can be suggested to be applied to attempt to predict travelling consumer behavioral changes in global online travelling business environment, the reasons are as below:

On the travelling consumer's beneficial hand, travelling consumers can apply this (AI) big data gathering method to attempt to gather any global airline e-tickets and/or travelling agent's package product data to be analyzed by this artificial intelligent learning system to compare human general marketing research method, e.g. survey, questionnaire, marketing plan etc. different human judgement methods to predict traveler consumption behavioral change model. Then, it analyzed all the different data to compare what are the range of the most reasonable e-ticket and/or travelling package online purchase history and sale in order to make more accurate prediction to future traveler change traveling consumption behavioral model in next month, or next half year or next year short term period traveling consumption change prediction. Thus, it seems that future AI tool can be attempted to apply to predict any industries price behavior, e.g. deciding what level of price is the attractive level to attract consumer in these industries, e.g. fuel, education, tourism, health, entertainment, etc. different product purchase. It can give more absolute price suggestion to any merchants to set their price change predict in order to increase many customer numbers to buy their products in every year, or every quarter every month, or month week, even every day etc. different sale period.

Reference

Backman and others "motivation is conceptually viewed as " a state of need, a condition that services as a driving force to display different kind of behavior toward certain types of activities, developing preferences, arriving at some expected satisfactory outcome.", 1995, p.15.

Fishbein & Ajzen, "The model based on the three constructs of attitude, subjective norm, and perceived behavioral control". 1975.

Hsu et al. "A tourist behavior model has been developed, called the expectation, motivation and attitude " (EMA) model ,2010.

ICT,WWW . "Switzerland has one of the highest population-to-computer ratio in Europe." Switzerland, 2005.

Jorea Ministry of Environment, " For South Korea environmental attitude is a major factor in decision making vis-a-vis the consumption of " green" food and services", Korea, 2015.

Korea Ministry Of Environment. Public Organizations spend 2.2 Trillon Korean Won To Purchase green Products in 2014; Ministry Of Environment: Sejoung, Korea, 2015.

Lind , Lohmann & Danielsson , United Nations Population Division, "Demographic change is said to be one of the important drivers for new trends in consumer traveling change behavior in most European countries". 2001.

Mayne, Lonnie. " Evolve of die in the age of the consumer". Entrepreneur, N.P. , 16 Apr. 2014. web of Oct. 2016.

Lee, D.; Kim, M. ; Lee, J. adoption of green electricity policies: Investigating the role of environmental attitudes via big data-driven search-queries. Energy policy 2016. 90, 187-201.

Lee, Terrence, " Tech in Asia-connecting Asia's startup system " Tech. in Asia- connecting Asia's startup ecosystem, N.p.,4 July 2016.

Weber & Bottorn "risky decision is as choices among alternatives that can be described by probability distributions over possible outcomes" , 1989, p.114.

What factors can influence travel behavioural consumption

Prediction travel behavioral consumption from traditional human's mind of tourism market research method

How to predict travel consumption? It is one question to any travel agents concern to use what methods which can predict how many numbers of travelers where who will choose to go to travel more accurately. I think that who can consider how to predict travel behavioral consumption from psychology view and computer science view both.

On the psychology view, It has evidence to support the relationship between self-identify threat and resistance to change travel behavior to any travelers, controlling for whose past travelling behavior, resistance to change if a psychological phenomenon of long standing interest in many applied branches of psychology. Past travelling behavior has been acknowledged as a predictor of future action. Such as travelling behavior that is experienced as successful is likely to be repeated and may lead to habitual patterns. Some psychologists differentiate habit between two concepts, such as goal oriented and automatic oriented both. Although repeated past travelling behavior is addition goal oriented and automatic oriented. Further non-deliberative nature of habit may make appeals to judge and to predict future individual traveler's behaviour accrately. However, repeated travelling behavior without a necessary constraint of goal orientation and automatic oriented both. So, it seems that psychological factor can influence any individual traveler why and how who choose to decide whose travelling behaviour.

On the computer statistic view, structural equation modeling is an extremely flexible linear-in-parameters multivariate statistical modeling technique. It has been used in modeling travel behavior and values since about 1980 year. It is a software method to handle a large number of variables, as well as unobserved variables specified as linear combinations (weighted averages) of the observed variable.

Whether climate change can influence travelling behaviours.

The flexibility of human travelling behavior is at least the result of one such mechanism, our ability to travel mentally in time and entertain potential future. Understanding of the impacts is holidays, particularly those involving travel. Using focus groups research to explores tourists' awareness of the impacts of travel own climate change, examines the extent to which climate change features in holiday travel decisions and identifies some of the barriers to the adoption of less carbon intensive tourism practices. The findings suggest many tourists don't consider climate change when planning their holidays. The failure of tourists to engage with the climate change to impact of holidays, combined with significant barriers to behavioral change, presents a considerable challenge in the tourism industry.

Tourism is a highly energy intensive industry and has only recently attracted attention as an important contributions to climate change through greenhouse gas emissions. It has been estimated that tourism contributes 5% of global carbon dioxide emissions. There have been a number of potential changes proposed for reducing the impact of air travel on climate change. These include technological changes, market based changes and behavioral changes. However, the role that climate change plays in the holiday and travel decisions of global tourists. How the global tourists of the impacts travel has on climate change to establish the extent to which climate change, considerations features in holiday travel decision making processes and to investigate the major barriers to global tourists adopting less carbon intensive travel practices. Whether tourists will aware the impacts that their holidays and travel have on climate changes.

When, it comes to understand indvidual traveler's behavioral change, wide range of conceptual theories have been developed, utilizing various social, psychological, subjective and objective variables in order to model travel consumption behavior. These theories of travel behavioral change operate at a number of different levels, including the individual level, the interpersonal level and community level. Whether pro-environmental behavior can be used to predict travel consumption behavior in a climate change. However, the question of what determines pro-environmental behavior in such a complex one that it can not be visualized through one single framework or diagram.

Despite the potentially high risk scenario for the tourism industry and the global environment, the tourism and climate change ought have close relationship. Whether what are the important factors and variables which can limit tourism? e.g. money, time, family problem, extreme hot or cold weather change, air ticket price, journey attraction etc. variable factors. Mention of holidays and travel were deliberately avoided in the recruitment process, so as not to create a connection factor to influence traveler's individual mind. However, the dismissal of alternative transportation modes can be conceived as either a structural barrier, in the sense that flying is perhaps the only realistic option to reach long-haul holiday destination, or a perceived behavioral control barriers in that an individual perceives flying as the only option open to whom. The transportation tool factor will be depend to extent on the distance to the destination. This can also be interpreted in a social perspective as an intention with the resources available where much international tourism is structured around flying. To increase the availability of different transportation modes, tourists could choose holiday destination closer to home.

Finally, also how to predict future travel behavioural consumption. I feel that travel agents need to predict whether any country's random daily variation of weather factor is also important to influence travel behaviour. e.g. in weather, temperature, rainfall adn snowfall with traffic accidents factors will have relationship to cause travel demand. Some scientists estimate suggest that when warmed temperatures and reduced snowfall are associated with a moderate decline in non-fatal accidents, they are also associated with a significant increase in fatal accidents. Thus increase in fatalities and temperature. Half of the estimated effect of temperature on fatalities is due to changes in the exposure to pedestrians, bicyclists and motorcyclists as temperature increase. So, if any countries have rainfall, snowfall and low temperature to cause traffic accidents, whether this accident occurrence will influence the travelers who liking climb snow hills, riding bicycle, running sports who will avoid to travel to these countries' bad weather after occurs. So, why I feel that this natural climate factor will also be one serious factor to influence travel behavioral consumption.

8.1 Market method predicts future travel consumption behavior

Whether individual habitual behaviour can influence travelling behaviour : e.g. renting travel transportation tools

Whether habit can be intended to predict of future travel behavior to people are creatures of habits. Many of human's everyday goal-directed behaviors are performed in a habitual fashion, the transportation made and route one takes to work, one's choice of breakfast. Habits are formed when using the some behavior frequently and a similar consistency in a similar context for the some purpose whether the individual past travel consumption model will be caused a habit to whom. e.g. choosing whom travel agent to buy air ticket or traveling package; choosing the same or similar countries' destinations to go to travel ; choosing the business class or normal (general) class of quality airlines to catch planes. Does habitual rent traveling car tools use not lead to more resistance to change of travel mode? It has been argued that past behavior is the best predictor of future behavior to travel consumption. If individual traveler's past consumption behavior was always reasoned, then frequency of prior travel consumption behavior should only have an indirect link to the individual traveler's behavior. It seems that renting travel car tools to use is a habit example. So, a strong rent traveling car tools useful habit makes traveling mode choice. People with a strong renting of traveling car tools of habit should have low motivation to attend to gather any information about public transportation in their choice of travelling country for individual or family or friends members during their traveling journeys.

Even when persuasive communication changes the traveler whose attitudes and intention, in the case of individual traveler or family travelers with a strong renting travel car tools habit. It is difficult to change whose travel behaviors to choose to catch public transportation in whose any trips in any countries. However, understanding of travel behavior and the reasons for choosing one mode of transportation over another. The arguments for rent traveling car tools to use, including convenience, speed, comfort and individual freedom and well known. Increasingly, psychological factors include such as, perceptions, identity, social norms and habit are being used to understand travel mode choice. Whether how many travel consumers will choose to rent traveling car tools during their trips in any countries. It is difficult to estimate the numbers. As the average level of renting travel car tools of dependence or attitudes to certain travel package policies from travel agents. Instead different people must be treated in different ways because who are motivated in different ways and who are motivated by different travel package policies ways from travel agents.

In conclusion, the factors influence whose traveler's individual behavior either who chooses to rent traveling car tools or who chooses to catch public transportation when who individual goes to travel in alone trip or family trip. It include influence mode choice factors, such as social psychology factor and marketing on segmentation factor both to influence whose transportation choice of behavior in whose trip.

How to determine future travel behavior from past travel experience and perceptions of risk and safety for the benefits to travel consumers?

How to determine future travel behavior from past travel experience and perceptions of risk and safety for the benefits to travel consumers? Why does individual traveler avoid certain destination(s) is(are) as relevant to tourist decision making as why who chooses to travel to others. Perceptions of risk and safety and travel experience are likely to influence travel decisions. If travel agents had efforts to predict future travel behavior to guess whether travelers will feel where is(are) risk and unsafe to cause who does not choose to go to the country to travel. Then, the travel agents will avoid to choose to spend much time to design the different traveling package to attract their potential travel consumers to choose to travel. The reason is because in the case of individual traveler's tourism experience, the traveler whose past disappointment travel experience (psychological risk) will be a serious threat to the traveler's health or life (health, physical or terrorism risk). The past safety or unhealthy risk to the country(countries) will influence the traveler decides to choose not to go to the countries(country) to travel again in the future.

What is push and pull factors to influence any
traveler who chooses where is whose preferable travelling destination

How to predict individual traveler's behavioral intention of choosing a travel destination. Understanding why people travel and what factors influence their behavioral intention of choosing a travel destination is beneficial to tourism planning and marketing. In general, an individual's choice of a travel destination into two forces. The first force is the push factor that pushes an individual away from home and attempt to develop a general desire to go somewhere, without specifying where that may be. The other force is the pull factor that pull an individual toward in destination, due to a region-specific or perceived attractiveness of a destination. The respective push and pull factors illustrate that people travel because who are pushed by whose internal motives and pulled by external forced of a destination. However, the decision making process leading to the choice of a travel destination is a very complex process. For example, a Taiwanese traveler who might either choose new travel destination of Hong Kong or another old travel Asia destinations again or who also might choose any one of Western country, as a new travel destination. The travel agents can predict where who will have intention to choose to travel from whose past behavior and attitude, subjective and perceived behavioral control model.

The factors influence where is the traveler choice, include personal safety, scenic beauty, cultural interest, climate changing, transportation tools, friendliness of local people, price of trip, trip package service in hotels and restaurants, quality and variety of food and shopping facilities and services etc. needs. So, whose factors will influence where is the individual travel's choice. It seems every traveler whose choice of travel process, will include past behavior. e.g. travelling experience, travelling habit, then to choose the best seasoned travelling action to satisfy whose travel needs. This process is the individual traveler's psychological choice process, who must need time to gather information to compare concerning of different travel packages, destination scene, climate change, transportation tools available to the destination, air ticket price etc. these factors, then to judge where is the best right destination to travel in the right time.

Why expectation, motivation and attitude factor can influence travelling behaviour.

Social psychology is concerned with gaining insight into the psychological of socially relevant behaviors and the processes. For instance, on a global level bad influence to global warming, it influences some countries extreme cold or hot bad climate changing occurrence, then it ought influence some travelers' behavioral decision to change their mind to choose some countries to go to travel at the moment which do not occur extreme hot or cold climate (temperature). e.g. above than 40 degree in summer or below than 0 degree in winter. Due to the extreme climate changing environment in the countries, it will cause them to feel uncomfortable to play during their trips. So, the global warming causes to climate changing factor will influence the numbers of travel consumption to be reduced possibly. This is global climate changing environment factor influences to bad or uncomfortable social psychological feeling to global travelers' mind of traveling decision. What is individual traveler expectation, motivation and attitude? Tourism sector includes inbound (domestic) tourism and outbound (overseas) tourism both incomes to any countries. According to recent article, a tourist behavior model has been developed, called the expectation, motivation and attitude (EMA) model (Hsu et al., 2010).

This model focuses on the pre-visit stage of tourists by modeling the behavioral process by incorporating expectation, motivation and attitude. Travel motivation is considered as an essential component of the behavioral process, which has been increasing attention from the travel; industry. The economic approach defines "tourism" is an identifiable nationally important industry. It includes the component activities of transportation, accommodation, recreation, food and related service. So, tourism behavioral consumption is concerned the individual tourist's usual habituate of the industry which responds to whose needs, and of the impacts that both the tourist and the tourism industry have on the socio-cultural, economic and physical environment.

However, travel motivation means how to understand and predict factors that influence travel decision making. According to Backman and others (1995, p.15), motivation is conceptually viewed as " a state of need, a condition that services as a driving force to display different kind of behavior toward certain types of activities, developing preferences, arriving at some expected satisfactory outcome." So, motivation and expectancy which has close relationship to any tourist before who decided to do any tourism of behavior. Some economists confirmed motivation and expectancy which has relations, such as expectation of visiting an outbound destination has a direct effect on motivation to visit the destination; motivation has a direct effect on attitude toward visiting the destination; expectation of visiting the outbound destination has a direct affect on attitude toward visiting the destination and motivation has a mediating effect on the relationship in between expectation and attitude.

8.2 What methods can predict future travel behavioural consumption

How to use qualitative of travel behavioural method to predict future travel consumption.

I also suggest to use qualitative of travel behavioural method to predict future travel consumption. Methods such as focus groups interviews and participant observer techniques can be used with quantitative approaches on their own to fill the gaps left by quantitative techniques. These insights have contributed to the development of increasingly sophisticated models to forecast travel behavior and predict changes in behavior in response to change in the transportation system. First, survey methods restrict not only the question frame but the answer frame as well, anticipating the important issues and questions and the responses. However, these surveys methods are not well suited to exploratory areas of research where issues remain unidentified and the researched seek to answer the question “why?”. Second, data collection methods using traditional travel diaries or telephone recruitment can under represent certain segments of the population, particularly the older persons with little education, minorities and the poor. Before the survey, focus group for example can be used to identify what socio-demographic variables to include in the survey, how best to structure the diary, even what incentives will be most effective in increasing the response rate. After the survey, focus, focus groups can be used to build explanations for the survey results to identify the ”why” of the results as well as the implications. One Asia Pacific survey research result was made by tourism market investigation before. It indicated the travel in Asia Pacific market in the past, had often been undertaken in large groups through leisure package sold in bulk, or in large organized business groups, future travelers will be in smaller groups or alone, and for a much wider range of reasons. Significant new traveler segments, such as female business traveler. The small business traveler and the senior traveler, all of which have different aspirations and requirements from the travel experience.

Moreover, Asia tourism market will start to exist behaviors in the adoption of newer technologies, a giving the traveler new ways to manage the travel experience, creating new behaviors. This with provide new opportunities for travel providers. The use of mobile devices, smartphones, tablets etc. and social media are the obvious findings to become an integral part of the travel experience. Thus, quality method can attempt to predict Asia Pacific tourism market development in the future.

However, improving the predictive power of travel behavior models and to increase understanding travel behavior which lies in the use of panel data(repeated measures from the same individuals). Whereas, cross-sectional data only reveal inter-individual differences at one moment in time, panel data can reveal intra-individual changes over time. In effect, panel data are generally better suited to understand and predict (changes in) travel behavior. However, a substantial proportion was also observed to transition between very different activity/travel patterns over time, indicating that from one year to the next, many people renegotiated their activity/travel patterns.

How to apply advanced traveler information systems (ATIS) to predict future travelling behaviour.

Nowadays, information can impact on traveler behavior and network performance. For example, when steadily growing levels of vehicle ownership and vehicle miles traveled information has been identified as a potential strategy towards man aging travel demand, optimizing transportation networks and better utilizing available capacity. Toward, this goal to predict further tourist behavioral consumption. Many countries, government tourism development institutes has applied advanced traveler information systems (ATIS) which travel behavior models and high-fidelity network performance models made increasingly feasible through the rapid advances in computer power. Crucial components of this problem domain are the modeling of individual tourist drivers' response to travel information and the development accurate guidance of relevance to real would trip makers. So, this advanced traveler information systems (ATIS) can assist the tourist who like to rent travelling car tools to travel in any countries own free traveler information systems service conveniently. Also, this travel information system can be intended to assist travelers to make better travel choices. e.g. this system can improve the decision making of individual traveler rather than improvements of network performance overall. So, we need to understand how tourists make their travel plans. Also, understanding decision process that lead to booking of the trip is equally important, as it allows of a potential behavior.

How does online tourism sale channel can influence traveling consumption of behaviour.

Nowadays, internet is popular, it seems that booking air ticket behavior of using internet is predicted to influence overall tourism air tickets payment method. Tourism industry has grown in the previous several decades. Despite its global impact, questions related to better understanding of tourists and whose habits. Using online travel air ticket booking benefits include booking electronic air tickets can be made from entering any electronic travel agents websites in the short time and electronic travel ticket payers do not need leave home, who can pay visa card to pre booking any electronic travel ticket from online channel conveniently.

How to analyze activity based travel demand ? Nowadays, human are concerning the traffic congestion and air quality deterioration, the supply oriented focus of transportation planning has expanded to include how to manage travel demand within the available transportation supply. Consequently, there has been an increasing interest in travel demand management strategies, such as congestion pricing that attempts to change aggregate travel demand. The prediction aggregate level, long term travel demand to understanding disaggregate level (i.e. individual levels) behavioral responses to short term demand policies, such as ride sharing incentives, congestion pricing and employer based demand management schemes, alternate work schedules, telecommuting limitation of travel agent traditionally work nature shall influence oriented trip based travel modelling passenger travel demand indirectly.

Finally, online travel purchase will be popular to influence the number of travel behavioural consumption nowadays. Any travel package products can be sold from websites to attract travellers to choose to prebook air ticket for any trips conveniently. In the past ten years, the internet has become the predominant carrier of all types of information and transactions. Regarding travel decisions, internet has also become an important sales channels for the travel industry, because it is associated with comparably lower distribution and sales costs, but also because ir adapts to hign supply and demand dynamics in this industry. Consequently, the travel and tourism industry tries to increase the internet sale specific share of sales volumes. So, internet sale channel has changed travel consumption behavioural pattern and characteristics and travel experience. For example, Switzerland has one of the highest population-to-computer ratio in Europe. It is also one of the most highly internet penetrated countries in terms of use of the WWW on a day-to-day basis, with more than 75 percent of the population older than 14 years using the WWW daily (ICT, 2005).

The reason of booking online tourism may include: convenience, fast transaction, finding traveling package choice easily, more airline seats available. So, online booking tourism will influence the traditional tourism agents visiting of sales and air tickets and travelling package numbers to be decreased. Finally, the online booking tourism market shares will be expanded to more than traditional tourism agents visits sale market in the future one day. So, the travel agents who still use the traditional tourism visiting sale channel which ought raise whose features to compare to differ to online tourism sale channel if these traditional touriam agents want to keep competitive ability in tourism industry for long term.

Actively based patterns of urban population of travel behavioural prediction method.

Actively based patterns of urban population. It is a method of motivational framework means in which societal constraints and inherent individual motivations interact to shape activity participation patterns. It can be used to predict one city or urban the numbers of travel demand in the year. It has two elements: First, capability constraints refer to constraints are imposed by biological needs, such as eating and sleeping and/or resources, such as income, availability of cars etc. to undertake the urban or city's family activities in the year. Second, coupling constraints define where, when and the duration of planning activities that are to be pursued with other individuals. So, this method needs to gather information (data) to get the relationship between activities, travel and spending work time and space time to evaluate whether there are how many families who have real needs to spend time to go to travel in the year.

What is trip based versus activity based approaches?

What is trip based versus activity based approaches? The fundamental difference between the trip-based and activity based approaches is that the former approach directly focuses on trips without explicit recognition of the motivation or reason for the trips and travel. The activity based approach , on the other hand, views travel as a demand derived from the need to pursue travel activities. So, it is better understand the individual or family behavior basis for individual or family travelling decision regarding participation in travelling activities in certain places or cities or countries at given times and hence the resulting travel needs. This behavioral basis includes all the factors that influence the why, how, when and where of performed activities and resulting individuals and household, the cultural/social norms of the community and the travel surrounding environment.

Another difference between the two approaches is in the way travel is represented. The trip based approach represents travel as a collection of trips. Each trip is considered as independent of other trips, without considering the inter-relationship in the choice attributes , such as time, destination and mode of different trips. As tours are chains of trips beginning and ending at a same location , say home or work. The tour based representation helps maintain the consistency across and capture the interdependency and consistency of the modeled choice attributed among the trips of the same tour.

In addition to the tour based representation of travel, the activity based approach focuses on sequences or patterns of activity participation and travel behavior, using the whole day or longer periods of time is the unit of analysis. Such as approach can address travel demand management issues through an examination of how people modify their activity participation, for example, will individuals substitute more out-of-home activities for in home activities in the evening of who arrived early form work due-to a work schedule change?

The major difference between trip based and the activity based approaches is in the way, the time dimension of activities and travel is considered. In the trip based approach, time is reduced to being simply a cost making a trip and a day's viewed as a combination, defined peak and off peak time periods. On the other hand, activity based approach views individuals' activity travel patterns are a result of their time use decisions with a continuous time domain. As individuals have 24 hours in a day or multiples of 24 hours for longer periods of time and decide how to use that travel among or allocate that time to activities and travel and with who, subject to their socio-demographic, transportation system and other and scheduling of trips. So, determining the impact of travel demand management policies on time use behavior is an important step to assessing the impact of such policies on individual travel behavior. The final major difference between this two approaches relates to the level of aggregation. In the trip based approach, most aspect of travel, e.g. number of trips etc. are analyzed at an aggregate level.

Consequently, trip based methods accommodate the effect of socio-demographic attributes of households and individuals in a very limited fashion, which limits the activity of the method to evaluate travel impacts of long term socio-demographic characteristics of the individuals who actually make the activity travel choices and the travel service characteristics of the surrounding environment. So, the activity based models are better equipped to forecast the longer term changes in travel demand in response composition and the travel environment of urban areas. Also, using activity based models, the impact of policies can be assessed by predicting individual level behavioral responses instead of employing trip based statistical averages that are aggregated over defined demographic segments.

Why senior age will be main travelling target.

In the past, Germany government had established tourism survey analysis to analyze survey data in order to arrive at reliable conclusions on future trends in travel behavior. To aim to find how demographic change will influence the tourism market and how the industry can adapt to those changes. The travel analysis provided data on tourism consumer behavior, including attitudes, motives and intentions. Since, 1970 year, it is based on a random sample, representative for the population in private households aged 14 years or older. Then, a continuous high scientific standard combined with a national and international users makes the travel analysis a useful tool and reliable source for tourism industry and policy decisions. It aimed to gather statistical data. e.g. on the age structure and on demographic trends, quantitative and qualitative analysis with time series data from the travel analysis. It shows e.g. not only the future volume , quite different from today's seniors, or how who will travel of family holidays will change, e.g. single parents of low, but grandparents of growing significance for tourism.

Demographic change is said to be one of the important drivers for new trends in consumer traveling change behavior in most European countries (e.g. Lind 2001). Because the growing number of senior citizens in the European Union and other industralised countries, such as the USA and Japan, looks to become one of the major marketing challenges for the tourism industry. United Nations statistics predict that the share of people being 60 age or older will grow dramatically in the coming future, and is expected to rise from 10 percent of the world population in 2000 year to more than 20 percent in 2050 year (United Nations Population Division, 2001). From its statistic, some data showed that travel propensity increased throughout life until the age of about 50 years of age and was then kept stable until very late in life 75 age. The most important results is that the travel propensity when getting older is not going down between 65 and 75 age of course, the overall development of this variable is influenced by a lot of other factors which are rsponsible for quite a variation over time. It is now possible to suggest that the general pattern of travel propensity is one of the key indicators for holiday life cycle travel behaviour, includes three stages. The growth stage tends to increase from early aduithood until 45 age old or when reaching some 80%. The next stage is stabilisation from the ages of around 50 age,until 75 age old, starting with a lower increase. Finally, the decrease stage is a slight decrease occurs once people reach the more advanced age of 75 age to 85 age old (Lohmann & Danielsson 2001).

So, it seems Germany government tourism prediction to future travellers' behaviour indicated these findings, such as on how future senior generations will travel, who had used survey data to examine the patterns of travel behaviour of a generation getting older and applied the findings to draw conclusions on the future. Also, it predicted that on the future of family trips, family semgmentation will be the travel behaviour patterns in the future. These findings together with the statistical data on demographic change allowed for a better understanding of the coming tends in family holidays. It's aim developed in consumer behaviour related to demographic change and predicted what will happen future of tourism one had to consider other influences and drivers as well, for example, trends on the supply side. e.g. low cost airlines or in travelling consumption behaviour in general whether how the past may provide a key to predict travel patterns of senior sitizens to the future.

Given the projected growth of the senior citizens market, designing specific marketing strategies to meet the prospective needs of elderly tourists will become increasingly important. It has been an implict assumption that it will be a close relationship between the travel behaviour of today's senior citizens and the those of future ones. The growing number of senior citizens in the world. e.g. China, Hong Kong, Japan, USA etc. countries. Global senior citizen tourism market will be based solely on demographic predictions about the future of the population's age structure. However, many of these seniors won't only live longer but will be fitter and more active until later in life. Many of the will also have plenty in life. Many of them will also have plenty of time and money to spend on travel. So, will these new seniors behave like today's senior citizens? Will they adopt the same travel behaviour as the previous generation or become a new market of oldies for the leisure and tourism indudtry? However, to determine the actual number of senior citizens who will be travelling and to sought to evaluate and specify certain difficult to predict the actual numbers of senior citizen to any country. However, they can be based on the implicit assumption that there is a close relationship between the travel behaviour of past, present and future seniors. But is this a valid assumption? As the reiseanalyse travel analysis survey, which was conducted in Germany every year, offered some interesting data possibiltieis. It was designed to monitor the holiday travel behaviour, opinions and attitudes of Germans and has been carried out since 1970 year, questions in the questionnaire. Data are based on face to face interviews, with a representative sample of more than 7,500 repondents, the interviews being carried out in January each year. All results refer to the average for the defined generated, which ranges generally over ten years. The group of people then at the age of 60 to 69 age is described. This corresponds to the same generation ten years ago, when they had an age of 50 to 59 age. When this methodological approach is not necessarily very sophisticated, it does have the important advantages of being cost effective.

Psychological method to predict travel behavioural consumption.

On the psychological view point, I think individual traveler's character will have those kind of personal characteristics. First, simplicity searchers value above everything ease not transparency in their travel planning and holiday making, and are willing to avoid having to go through extensive research. Second, cultural purists use their travel as an opportunity to immerse themselves in an unfamiliar looking to break themselves entirely from their home lives and engage. Sincerely with a different way of living. Third, social capital seekers understand that to be well travelled is a personal quality, and their choices are shaped by their desire to take maximum of social reward from their travel. They will exploit the potential of digital media to enrich and inform their experiences, and structure their adventures always keeping in mind they are being watched by online audiences. Finally, reward hunters seek a return on the investment who make in their busy , high-achieving lives. Linked in part to the growing trend of wellness, including both physical and mental self improvement who seek truly extraordinary and often indulgent or luxurious' must have experiences.

Why needs to know the personal character of individual traveler's characteristics. Because if travel agents could feel which kinds of individual traveler's character, then who can predict which kind of travel package to design to them more easily. For example, how to determine future travel behaviour from past travel experience and perceptions of risk and safety? We need to concern that the influences of past international travel experience, types of risk associated with international travel and the overall degree of safety feeling during international travel on individual's travelling experiences likelihood of travelling to various geographic regions on their next international vacation trip or avoidance of those regions, due to perceived risk. Because individual traveler's experience of safety risk degree to the countries, it will influence who chooses to go to the countries/country to travel again.

Why travellers avoid certain destinations are as relevant decision making as why who choose to go to the country(countries) to travel. Perceptions of risk and safety and travel experiences are likely to influence travel decisions; efforts to predict future travel behaviour can benefit to individual tourist's decision making. As Weber & Bottorn (1989) defined risky decision is as "choices among alternatives that can be described by prodability distributions over possible outcomes" (p.114). Some psychologists judge subjective perceptions of physical reality, i.e. image of a particular tourist destination, whereas value judgement refers to the way individual rank destinations according to whose attributes. i.e. attractiveness, safety, risk etc. factors to form on overall image. So, if the individual traveler had unhappy and worried and unsafe experiences to go to where the place(country) to travel during whose vacation time before. Then, this negative travel experience will influence who is afraid to go to the place (country) to travel again. Risk of place, country, destination or region means the danger is relatively high to the place, ie. increasing in airplane accidents, crime or terrorist activity targeting citizens of potential traveler's nationality or the probability of occurrence is great , ie. recent occurrences involving travel regions/destinations under consideration or effective actions to control consequences exist. i.e. selecting safe regions and destinations, taking extra precautions when traveling to risky destinations. These risk factors will influence the individual traveler who chooses to cancel travel plan to go to the country again.

Another interesting research, how to predict behavioural intention of choosing a travel destination, which has focus of toursm research for years, but the complex decision making process leading to the choice of a travel destination has not been well researched. The planned behaviour model using its core constructs, attitude, subjective norm and perceived behavioural control, with the addition of the past behavioural variable on behavioural intention of choosing a travel destination.

Understanding why people travel and what factors influence their behavioural intention of choosing a travel destination is beneficial to tourism planning and marketing. Understanding travel motivation is the push and pull model. The idea of the push and pull model is the decomposition of an individual's choice of a travel destination into two forces. The first force is the push factor that pushes an indvidual away home and attempts to develop a general desire to go somewhere else, without specifying where that may be. The second force is the pull factor, that pulls on individual toward a destination, due to a region specific travel location or perceived attractiveness of a destination. The respective push and pull factors illustrate that people travel because who are pushed by their internal motives and pulled by external forces of a destination. Nevertheless, how push and pull factors guide people's attitude and how these attributes lead to behavioural intentions of choosing a travel destination have rarely been investigated. The decision making process leading to the choice of a travel destination is a very complex process. The planned behaviour model is as a research framework to predict the behavioural intention of choosing a travel destination. The model based on the three constructs of attitude, subjective norm, and perceived behavioural control (Fishbein & Ajzen, 1975).

In conclusion, the factors can influence travelers who decide to choose to travel the country, which include personal safety was perceived to the highest motivation factors among the important factors which include, scenic beauty, cultural interests, friendliness of local people, price of trip, services in hotels and restaurants, quality and variety of food and shopping facilities and services. The factors include both push and pull. Push factors include knowledge, prestige, and enhancement of human relationship etc., whereas, the most significant pull factors include high technologic image, expenditure and accessibility etc. For example, Japanese travelers visiting Hong Kong. Push factors are such as exploration dream fulfillment and pull factors are such as benefits sought, attractions and good climate city. It will be the factor of future travel patterns and motivations of sub-cultural and ethic groups for Japanese choice to go to Hong Kong travelling.

Bibliography

Backman, K., Backman, S., Uysal, M. And Sunshine, K. (1995). Event Tourism : An Examination Of Motivations And Activities. Festival Management And Event Tourism, 3(1), 15-24.

Fishbein, M., & Ajzen, Z. (1975). Belief, Attitude, Intention And Behaviour: An Introduction To Theory And Research, Boston: Addison Wesley.

Hsu, C.H.C., Cai , L.A., Li, M(2010). Expectation, Motivation And Attitude: A Tourist Behavioral Model. Journal Of Travel Research, 49(3), 282-296. http://dx.doi, org/10.1177/004728750 9349266.

ICT Information And Communication Technology Switzerland, 2005. ICT Fakten (ICT facts). Available from http://www.ictswitzerland.ch/de/ict%2fakten/factsfigures.asp(retrieved Dec.12, 2005) in German.

Lind, (2001): Befolkningen, Familjen, Livscykeln- Och Ekonomisk Tillvaxt. Institutet For Tillvaxtpo-litiska studier/ Vinnova/Nutek.

Lohmann, Martin (2001): The 31 st. Reiseanalyse-RA 2001. Tourism: vol. 49, no.1/2001;pp.65-67, Zagreb.

United Nations Population Division (2001). World Population Prospects: The 2000 year Revision, New York.

Weber E.U., & W, P.Bottom (1989). "Axiomatic Measures Of Perceived Risk: Some Tests And extensions." journal of behavioral decision making, 2 (2): 113-31.

How does MTR (Mass Train Railway) need to consider route design location of choice by (AI) marketing research survey method?

Nowadays, transportation and economic development have close relationship. Economic development stimulates transportation demand by increasing the numbers of workers commuting to and from work, customers traveling to and from services areas, and products being moving by lorries on the roads between products and customers. According to Bailey, Mokhtarian and Little (2008) indicated "transportation route is past of distinct development pattern or road network and mostly described by regular street patterns as an important factor of human existence, development and civilization. The route network combined with increased road transportation investment result in changed levels of conveniently reflected through cost benefit analysis, savings in travel time, and other benefits. "

These benefits are noticeable in increased catchment areas for services and facilities , shops, schools, offices, banks and leisure activities by transportation route design of location choice.

Why MTR underground train transportation needs to know passenger behavior by (AI) marketing research survey method?

Understanding individual passenger behavior is essential for the design MTR transportation, because who can choose to catch bus, taxi, tram, train ferry etc. different kinds of public transportation tools. Individual traveler who decides to catch which kinds of public transportation tools, it depends on whether the public transportation tool can provide real time travel information, liking link travel time schedule. So, any country's (MTR) mass transit railway transportation enterprises need to understand where it has terminal to give convenience to the local living areas of time travelers to choose to catch MTR easily. Although, MTR ticket fare is one factor to influence any passengers choice. But, those other factors can also influence them to choice. e.g. MTR any terminal location of convenience, short time travelling, none crowding in busy (peak) time, MTR platform waiting arrival time, none sudden MTR engineering machines broken accident events occurrence frequently etc. different factors, any one of these factors which can influence passengers who choose to catch MTR or other kinds of transportation tools.

Why route choice can influence passenger behavioral choice by (AI) marketing research survey method ?

Usually, the busy time passengers will regard the route choice as a coordination problem to influence them to choose to catch which kinds of transportation tools. The route choice is as an opportunity costs to influence any busy time passengers to decide to choose to catch which kind of transportation tool which is the best right choice in the right time among of them. In the short time, for example, it seems any busy time passengers will choose to catch bus to substitute MTR underground train transportation tool, due to who feels the bus can arrive any destinations to compare other kinds of transportation tools in the most short time. However even if the MTR can either charge cheaper ticket fare to sell full day or charge discount ticket fare to sell in the busy (peak) time to compare to bus fare. It is possible that the busy time passengers will still choose to catch bus, if between the bus terminal and the another bus terminal that distance is the shorter time route to spend time to arrive destination to compare between the MTR terminal to the another MTR terminal arrival time . Also, although the busy time passengers will feel to encounter traffic jam to influence sitting or waiting bus time to be longer time in possible and who also feel MTR can avoid traffic jam problem. However, usually any busy (peak) time passengers will feel the chance of traffic jam occurrence will be less. So, the short bus route choice is more potential factor to influence the busy (peak) time passengers still to choose bus to catch.

However, if anyone wants to investigate results of day-to-day route choice which can be transferred to more realistic environment. It is necessary to explore individual behavior in an interactive experimental set up to ensure busy (peak) time passenger transportation behavioral choice. For example, a passenger has a choice between a main road (M) and a side road (S) for travelling from (A) to (B). (M) is faster if (M) and (S) are chose by the same number of

passengers. So, this method can be researched whether MTR terminal station is located at the main road (M) or the side road (S) where is more suitable to accept to passengers generally.
Why trip time reliability and crowding factors can influence MTR passenger choice by (AI) marketing research survey method?

Other problem is MTR busy (peak) time's crowding in public transportation occurrence of MTR underground train transportation tool is becoming a growth to concern as MTR demand growth at a busy (peak) time. To capture the MTR passengers benefits with reduced crowding from improved MTR public transport service and image. It is necessary a identify the relevant dimensions of crowding that are meaningful measures of what crowding means to MTR passengers. Two main influences on MTR model choice that are growing in relevance are trip time reliability and crowding. It represents a benefit-cost framework. In fact, MTR passengers can be willing to pay more expensive ticket fare, it MTR can avoid crowding and short and the accurate arrival trip time between terminals is reliable to occur. How to measure of MTR crowding, e.g. weighting the gap between the busy time, the standard (i.e. objective) and the perceived (i.e. subjective) metrics. We are not in a position to definitely map the two dimensions, which is a crucial requirement for translating objective improvements into equivalent subjective gains that then can be applied, willingness to pay estimates MTR ticket fares to obtain the additional MTR passenger benefits of MTR public transportation investment to any terminal stations.

Because MTR crowding has a negative impact on passengers in terms of psychological on emotional distress. MTR passengers are willing to stand for up to 20 minutes of the service is fast and reliable usually. However crowding outweighed these benefits from a MTR passenger's perpective, experienced crowding leads a increased dissatisfaction. e.g. stress and less privacy during who needs to stand up in MTR. Due to there are no enough places to supply to them to stand up in MTR. If the MTR trip time was longer time between the passenger's terminals, who will feel more dissatisfaction and it will cause who feels whether who ought need to choose to catch other transportation tools to substitute MTR next time. e.g. bus, train, tram, ferry, taxi etc. So, from an operator's perspective, the MTR service frequency or MTR size is significantly influenced by the level of ridership, which sends a signal to respond if the monitored crowding level exceeds the benchmark standard in the busy time. e.g. in the morning time or at the night time, the students or employment people who need to go to schools or offices (working places). The locations of different places between MTR terminals and crowding are regarded as a key service attribute for MTR pubic transportation along with other factors, such as travelling time and reliability, e.g. service quality, none engineering machines are broken to cause MTR stops suddenly.

Given the increasing importance of crowding on both the disutility to existing MTR public transportation users and the influence to it. MTR passenger can choose to use either the MTR public transportation or other public transportation. It is timely to review the MTR current measures of crowding defined by transportation authorities. MTR operators ought evaluate whether they appropriately reflect MTR each traveler experiences and perceptions of crowding in busy (peak) time. I suggest that MTR needs to buy other underground trains to supply to the busy (peak) time passengers to let them have enough seats to sit down, so who do not need to stand up in any MTR underground trains when they catch MTR underground trains in busy time. It aims to let who are willingness to pay the estimation of reasonable ticket fares to compare the other kinds of transportation tools in the busy (peak) time.

How MTR can attract many passengers by (AI) survey method?

On the commuter departure time choice of any reference point researching hand, the departure time decisions of communters are of fundamental importance of peak period MTR traffic congestion. However, whether on the demand side, MTR underground train congestion relief measures, such as MTR ticket fare to every terminal station needs to be charged cheaper fare or discount fare in the peak (busy) time every day. To aim to attract many passengers to choose to catch MTR Underground train public transportation tools, substitute to choose other public transportation tools in the peak time.

Over the past decades, there have been very active research efforts in the departure time problem, both in econometric modeling and dynamic user equilibrium fields. Although, these works provide valuable insights into dynamic commuter decision making, they do not identify the commuters' response to gains and losses related to whole actual arrival time to reference points who may have relative. The applicability of the reference point hypothesis of prospect theory to the commuter's departure time decision making to obtain a better understanding of how departure time choice in MTR platform during their waiting underground train arrival time. However, every MTR underground train actual arrival time and deviation variables related to reference points (gains and losses) are the key factors in the departure time choice model. How the MTR underground train of every communter's daily departure time decision can be modelled when the reference point hypothesis of prospect theory. The MTR underground train's schedule delay is defined as the difference between the preferred arrival time (PAT) and the actual arrival time (AT) for a given MTR communter. In a daily MTR commute, a commuter in the indifference band actual arrival time is an essential feature of MTR schedule study. Two reference points are the earliest acceptable arrival time and the work starting time for a given MTR platform waiting passengers. In psychological view point, prospect theory proposes that the displeasure of a loss is perceived or greater than the pleasure of a gain of the same attitude and therefore, the value function is stronger for losses than gains.

To conclude, it seems that if MTR waiting passengers need not spend long time to wait underground train arrival in platform and it can provide seats to let them to sit down in the busy (peak) crowding time. It will make them to feel pleasure, even the MTR ticket fare is not fair and reasonable to charge higher fare to compare other kinds of public transportation tools fares. So the peak waiting time factor can influence the passengers to choose other kind of transportation tools to catch easily. Moreover, MTR's two reference points are the earliest role. Similarly a loss is observed when the MTR platform waiting commuter experiences or actual arrival time which is beyond that the MTR schedule time. Due to that a MTR waiting commuter is as an early side arrival of whose actual arrival time is earlier than whose preferred arrival time.

In general, passenger transportation choice consumption behavior is similar to alcohol choice consumption behavior. Because some passengers choose to catch the kind of transportation tool , it is habit cause. Such as some alcohol consumers who often drive the brand of alcohol , it is habit cause also.

Some consumption psychologists had attempt to research whether planned behavior can predict alcohol consumption. This research aims to quantify variables between theory of planned behavior variables and (i) intentions to consume alcohol in habit and (ii) reducing alcohol consumption reasons. They showed some drunken violence alcohol consumers who will reduce to consume much alcohol if who feel driving accident or causing death or violence behavior or alcohol poison causing non-health after who have consumed too much alcohol often. Thus, it is important to understand the psychological determinants of alcohol consumption.

A model of human behavior that has been extensively utilized to predict health-related behaviors, such as alcohol consumption in the theory planned (TPB; Ajzen, 1991). This model proposes that the most important determinant of behavior is a person's intention to perform the behavior. Three variables are identified as determinants intention, attitude, subjective norm and perceived behavioral control . Attitudes are an individual's positive or negative evaluation of performing the behavior. Subjective norms reflect an individual's perceptions of social approach or disapproval for performing the behavior. It represents an individual's perceptions of control over behavioral performance in the face of internal and external barriers. These results suggest the possibility that alcohol consumer behavior that are harmful to health, such as alcohol consumption, may yield different relationships when compared with results for behaviors that are beneficial to health. Specifically, individuals may wish to emphasis a lack of control over health risk behaviors, because these behaviors are not seen as socially , desirable and may need to be explained away be reference to external causes, such as peer pressure (De Visser & Mc Donnell, 2013).

It seems that the passenger will reduce times to catch the kind of transportation tool if who feels the kind of transportation is dangerous (not safe) , such as if the alcohol consumer feels the alcohol will cause not health to him/ her. Then, who will also reduce times to choose to buy the brand of alcohol to drink.

Hence, it seems fear feeling psychological factor can influence alcohol consumers to reduce alcohol consumption in these situations, such as what action is being considered (e.g. heavy episodic drinking and the action is located (e.g. driving car) and what is the time frame for the action(e.g. needs car), the staff is a company driver when needs to drive car every day. Thus, whose action will influence to reduce whose alcohol consumption. Although, who has drinking alcohol in habit, but because who is one company driver, who is fear to cause accident to hurt himself/ herself and whose staffs when who sit in whose company car together. So, who will choose to reduce to consume alcohol in possible. Hence, dangerous is one most factor to influence passengers who do not choose to catch it.

However, social mobile analysis is a good method to research what the main factors which can influence passengers to either choose to catch MTR or choose other transportation tools, such as bus, tram, train, ferry , taxi etc. Some consumption psychologists had using a data set involving on adults (26 couples) living in a community for over a year to find that social behavior measured via face-to-face interaction, call and SMS logs, which can be used to predict the spending behavior to explore diverse business because loyal customers and overspend. Their results showed the mobile phone bases social interaction patterns can provide more predictive power on spending behavior than personality based features. Interestingly , these consumption psychologists found that more social couples also tend to overspend. Obtaining such insights about couple level spending behavior via novel social-computing frameworks can be of vital importance to economists, marketing professional and policy markers.

The basic idea is that a person's attitudes and behaviors are influenced by several levels of society, such as culture, subculture, social classes, reference groups and face-to-face groups. Such as any passenger's transportation tools choice, which are also influenced by culture, subculture, social classes, near transportation tool place, transportation cost, transportation time schedule and transportation service etc. factors.

In Special finally, some consumption psychologists investigated whether the social behavior measured via face-to-face interactions, call and SMS logs can be used to predict the spending behavior for couples in terms of their propensity to explore diverse businesses, engage frequently with them and overspend. Their findings not only motivate in potentially new line of investigation into a spending behavior via mobile sensing , but also demonstrate the feasibility of passive (i.e. which don't require active user attention) method, for undertaking similar studies at a large scale in near future.

In recent years, mobile sensing and reality approaches have been used to understand multiple aspects of human behavior. Insights on a behavioral level (e.g. overspending, loyalty and diversity) have much longer term validity and can explain certain aspects of human behavior. To study spending behavior of couples, the consumption psychologists focus on three important behavior : exploration, loyalty and overspending behavior. The aim is to identify the couples that tend to explore diverse business, engages repeatedly and frequently with certain businesses and (or) spend higher amounts of income to them. For exploration, who calculated the diversity in vendors frequented. The exact method for calculating diversity scores is explained in the next section. For loyalty, who considered how frequently couples engage with their favorite businesses. Specifically, they calculated the percentage of transactions (out of a couple's total transactions), that were made at their top businesses. Lastly, to quantify overspending, they calculated the ratio of the amount of money spent by a couple to their self-declared discretionary spending budget. Hence MTR can use mobile to enquire what are the general expectation to any passengers in order to predict the reasons why who prefer to choose other transporation tools more accurate.

Main barriers influence artificial intelligence consumer behavioral prediction

In future, it is possible that these barriers will influence how to apply (AI technology) to predict consumer behavior in success. The barriers may include: Lacking of a (AI) digital data gathering vision and strategy, lacking of efficient workforce readiness, (AI) technology constraints., non reaching (AI) consumer behavioral prediction mature stage, time and money and resource constraints, law and regulations prohibition to develop (AI) consumer behavioral prediction bug data gather technology.

However, the recommendation of solutions to attack the barriers to influence artificial intelligence consumer behavioral prediction not success, it may include gaining employee buy in to participate and develop (AI) consumer behavioral prediction technology, making customer experience to a concern (AI) big data gather questionnaire investigation, providing compensation, training to employees in order to achieve (AI) consumer behavioral big data questionnaire investigation research digital technological goals and strategy, task senior leaders manage any (AI) digital big data gather technology changes, putting policies and (AI) big data gather digital technology in place to support a fully remote, flexible workforce in any (AI) digital big data gather questionnaires research projects, teaching all employees how to code/understand (AI) big data gather consumer behavioral prediction software development, appointing a chief (AI) officer to manage any (AI) big data gather customer behavioral prediction projects and automate everything and encourage customers to attempt experience to self-service and (AI) big data gather questionnaire research to earn beneficial consumption aim after they gave feedback to any (AI) digital questionnaire researches. So, in the future, the (AI) digital big data questionnaire researches can include these industries surveyed, such as automat m financial services, public healthcare, private healthcare, technology, telecoms, insurance, life sciences, manufacturing, media and entertainment , oil and gas, retail and consumer products etc.

Hence, in the future, any of these industries can attempt to apply (AI) digital big data gather technology to predict how and why consumer behaviors will change in order to avoid reducing consumer number threat occurrence.

10.1 (AI) digital data gather technology predicts food consumer behavior's main barriers

What are the main barriers to food industry? When the food manufacturer applies (AI) big data gather technology to predict food consumer behavior? The barriers include that the food manufacturer / provider needs to decide whether when the right time is applied to the right (AI) digital big data prediction tool channel to find the right food consumers to be chose to full food consumption satisfactory questionnaires, how to gather multi-class food consumption classifiers on real-world food consumers transactional data from the food sale domain consistently to show the critical numbers of different kinds of food items at which the predictive performance most accurate? So, any food manufacturer / provider's advanced in (AI) digital data gather warehousing and management technologies can provide that opportunities for food business to enhance long term relationship with the food providers' clients.

However, food industry's (AI) digital data gather aims to improve food customer product targeting, increase food customer loyalty and food purchase probability to the food supplier. To effective identify, understand and satisfy the needs of their food customers, the food suppliers need to develop the right (AI) digital questionnaire questions and find the right food customers to fill every right questions from every digital questionnaire at the right time through the right channel.

Above of all these, they will be the barriers when one food supplier expects its (AI) digital data gather questionnaires which can conclude the most accurate prediction concerns any kinds of consumer food product choices. So, such as (AI) digital data prediction model, it is needed to incorporate into the food market segmentation, food customer targeting, and food challenging decisions with the goal of maximizing the total food customer lifetime. For example, (AI) big data gather transaction data is reasonable and accurate for building predictive models. Transaction data can be electronically collected and readily made available for data mining in lot quantity at minimum extra costs.

Suggestion to apply (AI) prototypes of food customer profiles method to predict food customer behavioral changes. Prototypes of food customer profiles mean to be extracted from the discovered bins and multi-class classifies models are built using those prototypes. The learned models can than be used to predict the class of food customer profiles (e.g. restaurants, school canteens, supermarkets etc. food suppliers) based on their food purchases. The approach is validated on the case study of a food retail and food service company operating in food and beverages market.

So, a food customer profile, it is a description (AI) data gather tool will record every of food customer using available information, which help in understanding their background and food consumption behavior. (AI) data gather tool can well develop every food customer profile, every food customer data is essential in food market analysis as they aid food suppliers in saving time and money by highlighting the real potential food consumers whose needs are to be met rather a range of individuals.

So, (AI) data gather tool can record every food consumer profile and every can be factual or behavioral food consumption. A factual food customer profile consists of a set of characteristics for (AI) big data gather record, e.g. demographic information , such as food customer name, gender, birth date, when a behavioral food customer profile consists of what the food customer is actually doing and is usually derived from (AI) digital transactional data gather record.

So, (AI) big data gather record's every behavioral food consumer profile can be much stronger predictor of the future food supplier consumption choice actions of a food customer. Furthermore, the food supplier's (AI) all past food consumer information that make up demographically based all past food customer profiles are expensive to acquire when the information for the food suppliers' past every food consumer food consumption behaviors. Moreover, food customer profile can be recorded to make real food purchase every time. So, when the food supplier finds the past food consumer's record from (AI) big data gather tool. Then, it can make more accurate judgement whether past every food consumer has chose to buy its food to eat how many times every year in order to predict whether its every past food consumer will choose to buy its foods how many times next year in possible. If the next year, its every past food consumer's consumption time to the food supplier is less than its current year consumption time. Then, the food supplier can attempt to find whether what factors to cause the past food consumers do not choose to increase food purchase times to the food supplier in current year. The factors may be possible be the food supplier's food prices are raised, food quality or taste is poor, the different kinds of food supply is shortage challenge, the food supplier's consumers lose confidence to buy the food supplier's foods to eat, when (AI) big data gather tool can help the food suppliers to find what the main factors to cause the past food consumer number to be reduced in order to predict how future food consumers' behavioral changes will be influenced from the food supplier's competitors in the global food supply market. Hence, (AI) big data gather tool can help every food supplier to attempt to find what the main factors to case the food supplier's food consumer number to be reduced as well as it can help the food supplier to predict how the food supplier's potential (past not every purchase its any food consumers) food consumers who can be persuaded to choose to buy its foods to eat by learning what the main factors influence.

In conclusion, (AI) big data gather tool can help the food supplier to find what the main factors influence its past food consumers do not choose to buy its food more times or find what the main factors will attract its potential (not ever buying its foods consumes) food consumers to choose to buy the food supplier's foods to eat.

10.2 The challenges of (AI) big data gather shaping the future of retail for consumer industries

Another challenge of (AI) big data gather is that how to shape the consumer behavior to let business owner to feel or know oe predict. It means that how it express it's conclusion or opinion for every consumer behavior after it had gather all big data in any data gather period, e.g. three months, half year or one year consumer shopping model data gather period.

Because every kind of industry, consumers will continue to demand price and quality change , with a wide range of convenient fulfilment options among of different kinds of products or services supply. Overall, the (AI) big data gather procedure gives opinion concerns every time retail experience will become more exciting, simple and convenient, depending on the consumer's ever-changing needs. So, I believe that (AI) big data gather every conclusion or result will be different, due to consumer's price and quality demand will often change to every kind of product or service supply in retail industry. So, how to shape (AI) big data gathering's analytical conclusion or result more clear. I shall recommend organizations need to build great understanding of and a stronger connection to increasingly empowered consumers before they plan and implement how to apply (AI) big data gather tool to predict consumer behavior as below:

Firstly, (AI) is empowered by technology, the consumer is redefining value. The traditional measures of cost, choice and convenience are still relevant, but not control and experience are also important. Globally, consumers have access to more than 2 billion different products choice by a wide range of traditional competitors and dynamic new entrants, all experimenting with new business models and methods of client engagement.

As choice increases, loyalty becomes more difficult familiarity and the consumer becomes more empowered. Businesses will have no choice and constantly innovate and disrupt themselves by meeting new technologies of high standards and expectations of consumers. So, (AI) data gather tool will need to follow different target group of consumers' needs to follow their different kinds of product design or style choice preferable to gather data in order to conclude the different target groups of consumer behavior to give opinion more clear and accurate to let businessmen to understand more clear how its customers' behavioral choice trend in the future half month, even to two years period.

Secondly, businessmen need to adopt changing technologies rapidly. Technology will be the key driver of this retail industry. Industry participants will only success if they have a clear prediction to focus on how to using technology to increase the value added to consumers. They must , however, do so will I realistic assessment of their costs and benefits. Hence, (AI) big data gather technological tools will need to design to help them to gather data efficiently by these ways, such as the internet of things (IOT), artificial intelligence (AI) machine learning, augmented reality (AR)/virtual reality (VR), digital traceability. So, future (AI) big data gather tool are predicted to be most influential customer behavioral positive emotion changing tool for retail , due to their widespread applications , ability to drive efficiencies and impact on labor in order to impact consumer behavior changing effort from negative emotion to positive.

Thirdly, (AI) big data gather tool is an advanced data science of consumer behavior predictive tool. Businesses will have to bring the journey from simply collecting consumer data to using it to scale and systematize enhanced decision making across the entire value chain. When focused on their business goals, industry players should not lose sight of the impact that future capabilities and transformative business models may have on society.

However, (AI) big data gather tool will encounter these challenges when any business plans and implements to apply it to predict consumer behavior in retail industry. The challenges include that as below:

1. The high cost and difficulty of implementing new technologies . The (AI) big data gather tool needs capital and capabilities to be designed to implement to be applied to different retail industry users. so, expensive barriers to innovation, an organization and the skillsets of its people to support a new design of (AI) big data gather tool, highly digital technology may be required.

2. Slow pace of cultural change. Consumers need to adapt or accept (AI) new technology consumption model in the traditional retail industry. The rate of change is outpacing the ability of businesses to keep up. (AI) big data gather tool needs to be designed to adopt in new or evolved business model requires, in most cases, a new level of customer behavioral predictive machine operation will impact to influence any retail businesses' consumer behavioral changes at a minimum, an organization's structure, capabilities, culture and decision making. If the retail business expects to apply (AI) big data gather tool to predict how to change its consumer behaviors and how their consumption behaviors will tend to change in order to achieve to change their positive emotion from negative emotion before they choose to buy its product or consume its service in success.

10.3 Challenge to using (AI) neural networks to predict customer behavior from big data gather tool

(AI) big data gather tool will encounter the challenge: How can predict customer behavior be represented as sequential data describing the interactions of the customer with a company or an (AI) data gather system through the time, e.g. these interactions are items that the customer purchase or views ? So, every customer data gather , (AI) needs to spend time to analyze how and why to cause whose consumption behavioral choice. It is too difficult matter or judgement for (AI) learning. So, (AI) needs to spend time to learn how to analyze every customer's shopping behavior or actin in order to gather all different consumers' past shopping action information in order to help business owners to predict future its potential customer shopping behavior how to change more clear and accurate prediction.

(AI) big data gather tool needs to learn to know that how to judge every customer interaction likes purchases over time can be represented with sequential data. Sequential data has the main property that the order of the information is important. Many (AI) machine learning models are not suited for sequential data, as they consider each input sample independent from previous ones. Therefore, at the end of the sequence, (AI) big data gather learn machines need to keep in their internal state of every customer purchase data, kind of product or service, price , whole year consumption times form all previous inputs, making them suitable for this type of data.

However, consumer behavior can be represented as sequential data describing the interactions through the time. Examples of these interactions are the items that the user purchases or views. Therefore, the history of interactions can be modeled as sequential data, which has the particular trial that an incorporate a temporal aspect. For example, if a user buys a new mobile phone, who might purchase accessories for this mobile phone in the near future or it the user buys a electronic book or paper book , he might be interested in books by the same author. Therefore, to make accurate predictions is important to model this temporal aspect correctly. To solve this predictive challenge of consumers to buy the product. One count the number of purchased products of a particular category in the last N days, or the number of days since the last purchase.

So, the (AI) big data gather designers can attempt to produce a feature vector which can be fed into a machine learning algorithm such as " logistic regression" will be the main feature and function to any (AI) big data gather machine to learn how to apply this " logistic regression" function or feature to predict any customer behavioral change for any product purchase or service consumption to the (AI) predictive consumer behavioral business users. Every different kinds of product purchases or services consumption will be needed to design " different model of logistic regression" in order to follow the kind of business to predict whose consumer purchase or service consumption behavior to predict more accurate.

10.4 Challenges of artificial intelligence, algorithms technology and machine learning impact to consumption market

Markets have played a key role in providing individuals and businesses with the opportunity to gain from trade. If (AI) big data gather tool can predict how to change potential customer behavior in success. The challenges to consumers will face that the overall market consumption model will be dominated by the businessmen only. So, it is not fair or reasonable to consumers, because (AI) big data gather tool has controlled or dominated all consumers' minds and it has predicted how and why every kind of product or service consumer shopping model or consumption behaviors how will change.

It will bring this questions: How can market designers learn the characteristics necessary to set optimal, or at least better, reserve prices after they had gather all data to conclude the analytical results of their consumers behaviors how will change? How can market designers better learn the environments of their markets?

In response to these challenges, artificial intelligence (AI) and machine learning are important tools for market design. For example, retailers and marketplaces , such as eBay, Amazon and many others are mining their vast amounts of data to identity patterns that help them create better shopping experiences for their clients and increase the efficiency of their markets. By having better prediction tools, these and their companies can predict and better manage dynamic consumption market environments. The improved forecasting that (AI) and machine learning algorithms provide help marketplaces and retailers better anticipate consumer demand and producer supply as well as help target products and activities for segmented markets. Another important application of (AI) 's strength in improving forecasting to help markets operate more efficiently is in electricity market example. To operate efficiently, electricity marker makers can attempt to apply (AI) machine learning tool to follow every household family electricity consumers' past electricity consumption record to judge (predict) how it will be every family's forecasting in the year.

An inaccurate forecast in the electricity supply and demand that can dramatically affect electricity market bad supply outcomes causing high variance in electricity charge prices or worse, blackouts. By better predicting every family's electricity demand and supply , electricity market makers can better allocate power generation to the most efficient power sources and maintain a more reasonable electricity stable charge market. Any example is design market, the application of (AI) algorithms to market design are already widespread and diverse.

(AI) algorithms technology , it is a safe that (AI) will play a growing role in the design and implementation of market over a wide range of applications. The challenges are that how (AI) can guarantee accurate to predict when and why and how consumer behavioral changes to any retail industries. In fact, retailers will need to discover the value that (AI) can bring to what benefits to influence their customer behaviors.

In the future, (AI) will bring their benefits to influence customers to build positive emotions to any retailers in these aspects as below:

1. Future (AI) big data gather tool will be an area of compute science that deals with giving machines , the ability to seem like they have human intelligence. In short, it is the power of a machine to copy intelligent human behavior. For examaple, machine learning algorithms are being integrated into analytics and customer relationship management platforms to uncover information on how to better serve customers, chat bots have been incorporated into websites to provide immediate service to customers.

2. (AI) adoption continue to rise with chat bots taking the lead. Due to increasing ease of deployment , instant availability and improved quality, chat bots will become more and more common to manage customer service queries and to make intelligent purchase recommendations. Also, retailers can engage this kind of technology to answer continue questions and supplement customer support with chat-based shopping experience. So, (AI) and declines personalized, customized and localized experiences to customers.

(AI) will be applied across the entire retail product and service cycle, firm manufacturing to post-sale customer service interactions. Hence, retailers can use (AI) to its fullest potential will be also to influence purchases in the moment and anticipate future purchases, guiding shoppers towards the right products in a regular and highly personalized manner.

3. (AI) technology can rise the conscious customers. Customers are demanding an increased interest in the ethical practice of the brands they buy from. Todays, customers have a well-developed sense of what is solely intended to drive sales. This has lead to a rise in consumers ho make values based judgements about what to buy and

where to shop. These consumers believe their purchase habits have an impact on the world. To win customers, retailers need have good conscious to predict consumers' desire. Future, (AI) data gather technology will be a good consumer behavior predictive tool to predict about for years will now become customer expectations and will have drastically changed the path to purchase. So, (AI) data gather tool is the predictive consumer expectations tool on every interaction, they have these brands.

4. Future (AI) can be impacted to influence consumer behaviors by its potential to free up time, enhance, quality, and enhance personalization. The industries include: Healthcare industry can apply (AI) to support diagnosis by detecting variations in patient data, early identification of potential pandemics, imaging diagnostics; automat industry can apply (AI) to autonomous fleets to ride sharing, semi-autonomous features, such as driver assist, engine monitoring and predictive, autonomous maintenance; financial service industry can apply (AI) to design the suitable personalized financial planning, fraud detection and anti-money laundering and automation of customer operation; transportation and logistics industry can apply (AI) to autonomous trucking and delivery, traffic control and reduced congestion and enhanced security; technology, media and telecommunications industry can apply (AI) to search media, and recommendation, customized content creation and personalized marketing and advertising to attract retailers to promote; retail and consumer industry can apply (AI) to design personalized production, anticipating customer demand, , inventory and delivery management; energy industry can apply (AI) to read and record smart metering , more efficient grid operation and storage and predictive maintenance; manufacturing industry can apply (AI) to enhance monitoring and auto-correction of processes, supply chain and production optimization and on-demand production.

Hence, future (AI) technology will impact consumer technology when any retailers apply it to assist its manufacturing processes or product sale or service provision processes to satisfy consumers' needs, it means that it can help any retailers to influence positive emotion to consumers in their whole sale or consumption or purchase proccesses.

5. (AI) and machine learning technologies make it possible to capture, process, and inter data on a massive scale effectively , then any human being could ever do. For example, Criteo's creative technology " Kinetic design" can apply insights from 1.2 billion monthly impressions to select and optimize individual branded advertisements components according to each shopper's preference and intent. This ensures more personalization and visually inspiring on brand ads. resulting in up to 12% more sales for (AI) technology advertiser clients.

Moreover, advertisers can now engage and inspire shoppers on a more personal level, rendering custom ads. it real-time for every impression. So, designer continues to learn from each design's success to make ads. more and more effective over time. Furthermore, brands are increasingly using paid search on retail sites to draw attention to their products on the crowded online shelf, e.g. Google shopping is a key growth area's more users are engaging with shopping ads. and across the globe. Google shopping has become essential to retailers' marketing strategies, but is a difficult channel to apply its tool to be promoted effectively . Thus, future (AI) and machine -learning technologies can dramatically improve digital commerce performance application to apply (AI) and machine learning to digital consumer. So, future (AI) technology can be applied to digital commerce aspect, it will fall into the categories of pattern recognition, classification, prediction and consumer behavior.

In conclusion, the benefits of using (AI) in digital commerce include: improved efficiency in discovering the relationships between datasets over traditional methods, which require complex modeling and coding, improved accuracy for clearly defined processes that involve a lot of manual processing, ability to deal with a large emotion of data with many attributes, for example: customer behavior data, multichannel and multi-device data , complex product data and fraud detection, more accurate analysis, such as customer segmentation sentiment, analysis and personalization frequent algorithum refreshes, such as several times a day, to capture the changes in customer and market behavior.

Finally, however, a lot of types predictive consumption behavior around (AI), in particulars that driven by vendors claiming their solutions are (AI) , ready and can deliver dramatic improvements over existing technologies. Application leaders for digital commerce can be misled into believing that (AI) can solve all their problems, which is not true for n in-depth discussion of the (AI) consumers and market behavioral predictive tool and machine -learning technologies bot. Thus, (AI) prediction consumer behavioral technology can give beneficial quantitative analysis for forecasting in business and market especially in consumer behavior and in the consumer decision-making process (consumer choice model) more effectively and efficiently.

Is Artificial Intelligent the most effective and accurate consumer behavioral tool?

Is (AI) the best and the most effective and accurate consumer behavioral prediction tool to compare other kinds of consumer behavioral prediction tools? Nowadays, retailing competitions are serious businessmen often find different kinds of methods to attempt to predict consumer changes. The consumer behavioral predictive methods can include as these below methods, instead of (AI) big data gathering tool.

Firstly, statistics is the popular mathematic method, it applies auto-regression, liner regression, structural equation modelling, logistic regression statistic techniques to be used to predict consumer behaviors. Secondly, it is classification method, it sis a support vector machine to assist businessmen to make consumer behavioral prediction, it also includes decision making tress diagram technique. Thirdly, it is rule mining method, it is algorithm, market base analytic etc. business marketing concept analytical tool, it also includes graph mining technique tool. Next, it is psychological prediction model tool, it is psychology prediction model too, it is a kind of psychological method to predict consumer behaviors. Finally, it is the most updated and potential artificial neural network (ANN) machine tool, it gathered big data, then it will carry on analyzing and applies psychological method to conclude the most accurate and reasonable solutions to give recommendation to businesses to predict when and how and why their consumer behaviors will change. So, it is one owned human mind's machine and owned psychological and analytical efforts to replace humans to make any judgement in order to make the most accurate predictive behavioral changes for consumers, instead of the traditional marketing concept and psychological and mathematic methods to predict consumer behavior, (AI) big data gathering tool will be another new tool.

What are the advantages of (AI) tool to be used to predict consumer behaviors as well as what are the different between it and other traditional consumer behavioral predictive tools? I shall explain as below:

Firstly, as above all case studies are explained to (AI) questionnaire design method benefit, I believe (AI) big data gathering tool can be applied to help human to analyze and design any the suitable valid questions to enquire any kinds of business consumers in order to gather the most meaning and useful opinions to conclude the most accurate consumer behavioral prediction for every questionnaire. So, future (AI)'s analytical effort and decision making effort most be exceed above human's judgement efforts. So, future (AI) can help human to design the most useful and meaning different kinds of valid questionnaire (survey) questions as well as assist humans to analyze and make accurate decision making and conclusions to give opinions to help businessmen to predict when consumer behaviors will change and how their consumption behaviors will change to influence their businesses in order to help them to make any efficient and effective and accurate solutions to avoid consumer number to be decreased and the most important benefit is that it can give opinions to help businessmen to explain why (what the factors) cause their consumer behaviors change suddenly. It will be human's efforts can not achieve to exceed (AI)'s efforts in the future.

Secondly, (AI) can make artificial machine judgement and analytical effort, without human misleading or unfair or unreasonable judgement. So, it can make more fair and reasonable and accurate conclusion to give opinions to predict when, how and why consumer behaviors will change suddenly to the kind of business in customer model building process and evaluating the results of customer relationship management –related investment more accurate.

Furthermore, (AI) big data gathering tool will help businesses to improve the success rate of acquiring customers, increasing sales and establishing competitiveness. (AI) big data gathering tool can give opinions how to build customer loyalty to be positive emotion impact and it can find solutions to avoid every client's negative emotion causes to bring complaints behavior to the businessman's product or service. For example, Telecom industry and aggressive research has been conducted in this by applying various data mining techniques to avoid long distance phone call users' complaints. If gathered any long distance phone call users' past complaint data to record what are their general complaint issues. Then, (AI) tool will analyze all these past complaint issues to conclude and give opinions to let Telecom knows whether which aspects encounter challenge that Telecom needs to improve it's long distance phone call services or functions in order to satisfy Telecom's long distance phone call users' needs for long term. After Telecom attempted to improve its services and/ or functions from (AI) opinions and solution methods, when it fell it's long distance phone call users have positive emotions to satisfy its service performance and function performance. Then, it can prove (AI) tool's opinions and solutions are useful. The consequence is that their complain numbers will be decreased and they won't plan to choose another long distance phone call telephone service company to replace Telecom long distance phone call service more easily.

So, (AI) big data gathering tool can concentrate on finding focus on components of customer relationship management method and datasets more accurate and efficient and effective than human's data gathering and analytical effort. It implies (AI) big data gathering tool has unique more efficient and effective and accurate dataset gathering and analytical and judgement and decision making effort, it is human can not achieve.

Thirdly, (AI) big data gathering tool has much customer loyalty predictive effort. It's effort is more easily subsequently selected, reviewed and classified to compare human's gathering data effort in whole data gathering and analytical process.

In (AI) big data gathering process, (AI) can organize whole big data gathering process and technique more easily in short time. It will include these four steps. The first stage is that customer identification stage, customer identification also known as acquisition has to do with targeting the population , who are most likely to become customer segmentation. So, (AI) can help different kinds of businesses to gather their competitors' consumer purchase behavior data in short time, it is human can not achieve. The second stage is that customer attraction stage, after (AI) maker has been segmented for the business when it has ensured to gather the businessman's global competitors' consumers data. Then it analyze these all data to find solutions / methods to give the best opinions to the organizations how to achieve the direct effort and resources into attracting the target customer segments. The third stage is that customer retention, it can be defined as the activity that an organization undertakes in order to reduce customer defections. TO be successful, customer retention starts with the first contact on organization has with a customer and continues throughout the entire lifetime of a relationship involves loyalty programs, one to one marketing and complaints management. SO, (AI) can consist the business to find the best or the most reasonable , efficient , effective solutions or methods and it will conclude all these solutions to find the most reasonable and useful opinions to achieve to the aim to help the business to reduce customer complain numbers and help the business to build confident loyalty relationship between it and its clients. SO, (AI)'s analytical effort and decision making effort can be more accurate than human's analytical effort and decision making effort. IT can achieve it's consumer behavioral predictive aim more accurate and efficient and effective in the shortest time to compare human.

Fourthly, (AI) big data gathering tool can design more accurate dataset program for questionnaire (survey) to compare human's questionnaire (survey) effort. It means that (AI) can spend less time to research and make judgement what are the most reasonable and meaning questions for different kinds of businesses' needs. This includes data conduction a questionnaire, survey or interview of the individual or environment researched, public data repository: This includes commercially available public data; organizational data; this contains data collected from an organizational database, organizational information system. For example, their website log details etc. It also includes company transactional data, data purchased from a company.

For example, one vehicle sale company expects to research all global vehicle sale companies' past the different kinds of vehicle styles, design sale number data, the different kinds of vehicle style, design sale price data, every country's vehicle consumer number to the vehicle purchase number data to the vehicle company in short time. (AI) big data gathering tool can help the vehicle sale company to gather all any one for these global vehicle sale competitors' past data in the short time. It is human effort, who can not achieve this efficient, effective and accurate data gathering aim for this vehicle sale company. Even, when (AI) had gathered all global it's vehicle competitors' past sale data, (AI) can make more accurate analytical and judgement and decision making effort to design different kinds of questionnaire (survey) questions to prepare to enquire it's different target segmentation vehicle potential clients in order to predict what are their needs to choose to buy any vehicles from the vehicle company. SO, (AI) tool can conclude more accurate conclusions and give the most reasonable and useful opinions to let the vehicle company to know in order to predict what are it's potential vehicle buyer's needs and manufacture the suitable vehicle styles or designs to raise their vehicle purchase desires.

Fifthly, (AI) tool is only one perfect tool for big data gathering in order to achieve accurate results and increased profit. What is (AI) big data gathering mean? The term " big data"gathering describes the accumulation and analytical of vast amounts of information, but big data is much more than a big amount of data. It is also the ability to extract meaning to sort through big volumes of numbers and find the hidden patterns, unexpected correlations and surprising connections that can be used in different industries like medical field, security and protection field or marketing that adopt " big data driven" decision making enjoy significantly greater productivity than those that do not. So, the benefits of (AI) is given to the company by using big data repaid complexity of implementation projects and hence project risks, when accelerating time to value. It is why that human's gathering effort can not replace (A I) data gathering effort.

All analysing above benefits to (AI) big data benefits to any organizations, it brinfs this question: How can (AI)apply big data gathering and analyzing to predict when and how any why consumer behavior will change suddenly? The purchase decision making process is consumers reducing purchase choice behaviors.

Consumers are being considered pure rational beings (consumer tried only to satisfy self-interest). Hence, due to future (AI) owns human's psychological , analytical , emotional predictive, purchasing decision making effort.

(A I) will be assumed to sees one customer how who will make purchase decisions. So, after the (AI) gathered all data concerns the find of business's past customer segmentation purchase activities, e.g. age, sex,. Income level, the product's style sale number, the product price variable sale etc. different kinds complex data.

It can makemore accurate psychological and analytical effort to predict when the business's consumer behaviors will change as behaviors will change as well as find what reasons their consumption behaviors will change and how trend of their consumer behaviors will change more accurate. For example, today there are a lot of industries that use big data: healthcare (treatment) becoming personalized and patient centric and predictive analysis are used to prevent diseases for example Angelina Jolie underevent a predictive double mastectomy after learning she had 87% rich to developing breast cancer, sports (by using sensors data are collected from players during a game in order to improve their playing schemes), weather(more than 60 years of global weather analysis are used to predict the risk of future extreme events), logistics (smart tucks and smart species, agriculture (monitoring weather and soil conditions for optimum point of harvesting).

Consequently, due to the evolving consumer demands, and the ever growing digitization, the world is digitally transforming which means the new technologies are needed to be used and driven significant business improvement. So, such as why (AI) tool will be our future main predictive tool to help businesses to predict when, how and why their potential customer behavioral will change. Big data is one of the our channels through digital transformation is made, together with cloud, mobile and networks. The challenges for digital transforming and therefore using A I big data gathering tool as main technology are: digital proficiency, legacy systems, security and jobs becoming absolute.

In the future, big data can use data from text to picture , sounds, movies, musics satellite coordinates or any other type of input or output data that type of input or uouput data that came from different influential aspect. It is cloud solutions, bring big data will be for predict insight driven by business stategy, new product strategies and new consumer relationship, predictive consumer behavioral strategy. Using the right data in the right business decision will mean smart decisions, new opportunitites and utimately a big competitive advantage.Hence (AI) big data gathering tool is different is that (AI) can be one depth in-memeory database function, it can make real-time data analytics that provide meaningful information in short time, it is also the visualization tool , such as SAP Lumira, allow this exploration and understanding of the data, and ultimately supports the decision making process. All above these features, which will be human's data gathering effort who won't exceed (AI) big data gathering effort. Hence, future (AI)big data gathering will be the best choice to assist businesses to predict consumer behaviors successfully.

Reference

Adrian, P. (2012). Introduction to marketing theory & practice,
3 rd edition, London: Oxford press.

Ajzen, I (1991). The theory of planned behavior. Organizational behavior and human decision processes, 50(2), 179-211. doi: 10.1016/0749.5978 (91) 90020-7.

Alba, Joseph W. and J. Wesley Hutchinson (1987). " Dimensions Of Consumer Expertise", Journal of consumer research, 13 March, 411-454.

Bailey, L., Mokhtarian, P.L. Little, A. (2008). The broader Connection Between Public Transportation, Energy Conservation And Greenhouse Gas Reduction, Report Prepared As Part Of TCRP Project J-11/Tasks Transit Cooperative Research Program, Transportation Research Board Submitted To American Public Transportation Association in http://www.apta.com/research/into/online/land_use.cfmi, accessed 17 April 2008.

Baucer, R,"Consumer Bhavior As Risk Taking , In Risk Taking And Information handling In Consumer Behavior", D. Coxceds Harvard University Press, Cambridge, Mass 1976.

Biederman, P. (2008). Travel and tourism, Pearson Prentice Hall, New Jersey.

Bogers, R. P., Brug, J. Van Assema, P., & Dagnetie, P.C. (2004) , Explaining fruit and vegetable consumption: The theory of planned behavior and misconception of personal intake level. Appetite, 42,157-166.

Bolton, Ruth N. (1998), " A Dynamic Model Of The Duration Of The Customer's Relationship With A Continuous Service Provider: The Role Of Satisfaction", Marketing Science, 17 (1), 45-65.

B.Shiv and A. Fedorikhin, " Heart And Min In Conflict: The Interplay Of affect And Cognition In Consumer Decision Making", J. Consumer Res., vol. 26, pp. 278-292, Dec. 1999.

Brown, K.W., Ryan, R.M. Reswell , J.D. (2007). Mindfulness: Theoretical Foundatins And Evidence For Its Salutary Effects. Psychological Inquiry, 18, 211-237.

Burke, R.R. : Behavioral effects of digital signage, J. Advertising Res. 49(2), 180-185 (2009).

Cant, M., Brink , A. & Brijall, S., Consumer behavior, Cape Town, South Africa: Juta, 2006.

Conner, M. & Abraham, C. (2001). Conscientiousness and the theory of planned behavior: Toward a more complete model of the antecedents of intention and behavior. Social psychology bulletin, 27, 1547-1561.

Cooper C. Mallon, K, Leadbetter S, Pollack L, Peipins (2005) , cancer internet search activity on a major search engine, United States 2001 to 2003, J Med Internet Res. 7(3): e36.

Cope, R. R. Cope and H. Davis (2008). Disney's virtual Queues: A strategic opportunity to co-brand services ? Journal of Business & economics research, vol. 6 no10, 13-20.

Cornelia, B.F. (1999) Rural development news, the North Central Regional Center For Rural Development vol. no 24 , IOWA.

Couper, M.P. J. Blair and T. Triplet (1999). A Comparison Of Mail And E-mail For a Survey Of Employees In USA Statistical Agencies. Journal Of Official Statistics, 15, 39-56.

David J. Nowak & Gordon M. Melsler (2016) " Air quality effects of urban trees and parks." National recreation and park association, USA.

Data monitor (2008). The proctor and gamble company. Retrieved Nov. 15 2009 from http://www.datamonitor.com/

De Hollander, A. E. M., J.M. Melse, Elebret & P. G.N. Kramers (1999), " An Aggregate public health indicator to represent the impact of multiple environmental exposures" Epidemiology: 606-617.

De Visser, R.O., & McDonnell, E.J. (2013). " Man points": Masculine capital and young men's health. Health psychology, 32(1), 5-14. doi:10. 1037/a0029045.

Dunn, J & A Neumsister (2002). Knowledge management in the Information age. E. business review, Fall , 37-45. Jounral of service, spring 2011, vol. 4, no1, De Grovte (2009).

Dyer, D., F. Dalzell & R. Olegario (2004). Rising tide. Lessons learned from 165 years of brand building at Procter and Gamble. Boston, MA: Havard Business School Press.

Eysenbach G (2006) Infodemiology: Tracking flu- related searches on the web for syndromic surveillance. American Medical Informatics Associaion Annual Symposium Proceedings , Curran Associates, Red Hook, NY, pp. 244-248.

Ettredge M, Gerdes, J. Karuga , G (2005) Using web- based search data to predict macro-economic statistics. Commun ACM 48: 87-92.

Felce, D. and Perry, J. (1995). Quality of life: A contribution to its definition and measurement, vol. 16, no.1 pp: 51-74.

Feldman, Jack M. And John G. Lynch Jr. (1988), "Self-
Generated Validity And Other Effects Of Measurement On Belife, Attitude, Intention And Behavior", Journal of applied psychology, 73(3),421-35.

Fiese, M, Hofmann, W., & Wanke, M (2009). The impulsive consumer. Predicting consumer behavior with implicit reaction time measurement. In M. Wanke (ed.) Social psychology of consumer behavior (pp.335-364). New York, NY: Psychology press.

Fitzsimons, Gavan, J. And Vicki G. Morwitz (1996), " The Effect Of Measuring Intent On Brand-Level Purchase Behavior", Journal of consumer research, 23 (1), 1-11.

Hallerman , D. (2008) video Advertising Online: Spending And Pricing , New York. E-Marketer.

Harriet Griffey. (2010) The art of concentration, enhance focus, Reduce, stress and achieve move. Macmillan publishers ltd,Basinastoke and Oxford, London UK.

Helleman, D. (2008) Video Advertising Online: Spending And Pricing , New York, E-Marketer.

Hensen, C. (2003). Kreuzfahrtourismus.www.christoph- hensen.de/Facharbeit.pdf.

Huang, H.I. (2012). An empirical analysis of the strategic Management of competitive advantage: a case study of higher technical and vocational education in Taiwan (Doctoral dissertation, Victoria University).

Jamieson, Linda F. And Frank M. Bass (1989), " Adjusting Stated Intention Measures To Predict Trial Purchase Of New Products: A Comparison Of Models And Methods," Journal of marketing research, 26 (August), 336-45.

Korea Ministry Of Environment. Public Organizations spend 2.2 Trillon Korean Won To Purchase green Products in 2014; Ministry Of Environment: Sejoung, Korea, 2015.

Kremers, S.P. J., De Bruijn, G.J., droomers, M., Van Lenthe, F. J., & Brug, J. (2005). Environmental interventions for selected dietary behaviors in adults. In J. Brug & F. J. Van Lenthe (eds.) , Environmental determinants and interventions for physical activity, nutrition and smoking: A review pp. 282-315. Rotterdam: Erasmus Medical Center.

Lee, D.; Kim, M. ; Lee, J. adoption of green electricity policies: Investigating the role of environmental attitudes via big data-driven search-queries. Energy policy 2016. 90, 187-201.

Lee, Terrence, " Tech in Asia-connecting Asia's startup system " Tech. in Asia- connecting Asia's startup ecosystem, N.p.,4 July 2016.

Los Angeles Country Department Of public Health (2016), Country Health Ranking Model, Retrieved From www.countryhealthrankgings.org/our-approach. USA.

Mayne, Lonnie. " Evolve of die in the age of the consumer". Entrepreneur, N.P. , 16 Apr. 2014. web of Oct. 2016.

McGregor, S.L. T., & Goldsmith, E.B. (1998). Expanding our understanding of quality of life, standard of living and well-being. Journal of family and consumer science, 90(2), 2-6, 22.

McMichael, A.J. M. Mckee, J. Shkolnikov and T. Valkanen (2004), " Morality trends and setbacks, global convergence or divergence?", Lancet 363, 1155-1159.

Melse, J.M. & A.E. M. De Hollander (2001). " Human Health And The Environment", background document for the OECD Environmental Outlook, OECD, Paris.

Moschis, George p. & Roy, L. Moore (1979), " Decision making among the young. A socialization perspective " Journal of consumer research , 6 (September).

Mulligan, M. Banerjee, T & Thomas, N. (2008) ,European Paid Content And Activity Forecast, (2008 to 2013), Jupiter Research.

Peter, J., Ryan, M, M, " An Investigation Of Perceived Risk At The Brand Level, " Journal of marketing research, 13 May 1976, pp. 184-188.

Pieters, R., & Wedel, M. (2007). Goal Control Of Visual Attention To Advertising: The Yarbus Implication. Journal Of Consumer Research, 34, 224-233 (August).

Parasuaman, and Leonard L. Berry (1985), " Problems And Strategies In Sevices Marketing", Journal of marketing, 49 (Spring), 33-46.

Priesnitz, W. (2007) Counting Our Food Miles. Natural Life, 1 July.

R.C. Oliver, " When is consumer loyalty?" J.Marketing vol. 63, pp.33-44.1999.

Reggiani, A . (ed). 1998, accessibility, trade and locational behavior, Ashgate publishing ltd, England.

Rushe, D. (2013) " The 10 best paid CEO in America". The Guardian , 22 Oct, (online). Available at: http://www.theguardian.com/business/2013/Oct22/best-paid-chief-executives-america (Accessed: 3 May 2014).

Spiekermann and Wegener (2007), update of selected potential accessibility indicators. Final report, urban and regional research (S&W), RRG spatial planning and geoinformation. ESPON. Available online at http:// <www.espon.eu/mmp/online/website/ contentprojects/947/1297/file_2724/ espon_accessibility_update-2006-fr_070207.pdf>, accessed on 1 July 2009.

Starbucks (2014) Our company available at http:// www. starbucks.com/about- us/company-information (accessed: 3 May 2014).

Shostack, G. Lynn (1984), " Designing Services That Deliver", Harvard Business Review, 62 (January-February), 133-9.

Shostack, G. Lynn (1985), " Planning The Service Encounter ,in the service encounter" , John A. Czepiel, Michael R. Solomon, and Carol F. Suprenant, eds. New York: Lexington Books, 243-54.

Shostack, G. Lynn (1987), " Service Positioning Through, Structural Change", Journal of marketing, 51 (Janurary), 34-43.

Soloman, Michael R. (1985), "Packaging The Service Provider", Service Industries Journal , 5(1), 64-71.

Stevens, C.W. (1980), "K-MartStores Try New Look To Invite More Spending" The Wall Street Journal, Nov. 26, 29-35.

Sullivan, Nicholas P(2007). You can hear me now: How Micro loans and cell phones are connecting the world, San Francisco, CA: John Wilsey & Sans, 2007.

T. Ambler, A. Ioannides, And S. Rose, " Brand s On The Brain : Neuroimages Of Advertising ", Business Strategy rev., vol. 11, 3. pp. 17-30. 2000.

Westbrook, Robert A. (1980), " Intrapersonal affective influences on consumer satisfaction with products, " Journal of consumer research , 7 (June) 49-54.

Wiig, k.(1993). Knowledge management foundations: Thinking About thinking. How people and organizations create, represent and use knowledge vol.1 , of knowledge management series schema press: Arlington, TX.

World Health Organization (2003). Diet, nutrition and the prevention of Chronic diseases report of a joint WHO/ FAO. expert consultation. Geneva: World Health Organization.

Wysocki, B. (1979), " Sight, Smell, Sound: They're all arms in retailer's arsenal" The Wall Street Journal, Nov. 17, 1979. 1-35.

Yale Center For Environmental Law And Policy (2006). Environmental Performance Index. Data available on-line at http://epi.yale.edu

TWO

(AI) -DRIVEN AUTOMATION INDUSTRY DEVELOPMENT

1.1 (AI) - driven automation industry development how to influence work nature change

(AI) -driven automation industry will create wealth and expand economy growth to any countries, but it will be accompanied by changed in the skills that workers need to learn. One of main ways that technology increases productivity is by decreasing the number of labor hours needed to create a unit of output. It implies (AI) technology will influence low educated and low skillful labor number to be decreased (reduction employment number).

In contrast, technological change tended to work in a different direction throughout the nowadays. The advance of computer and the internet raised the relative productivity of higher skilled workers. So, routine-intensive occupations that focused on predictable tasks disappearance, such as switch board, operators, filming checkers, travel agents and assembling line workers etc. were particularly replaced by new technologies.

However, today, it may be challenging to predict exactly which jobs will be most immediately affected by (AI) driven-automation. The reason is because (AI) is not a single technology, but rather a collection of technologies that are felt unevenly through the economy to influence job changing both negatively and positively. In positively view point, (AI) driven-automation will make many workers more productive and increase demand for certain skills. Consequently, new jobs are likely to be directly create in areas , such as the development and supervision of (AI) as well as indirectly created in a range of areas throughout the economy as higher incomes lead to expanded demand. Otherwise, in negatively view point, many traditional human needed (demand) skillful jobs will be threatened by automation are highly concentrated among lower-paid, lower-skilled and less -educated workers. It means automation will cause pressure on demand for this group, pressure and employment, if (AI) can replace the low skilled and less educated workers' jobs. Thus, (AI) will have negative influence to impact on the labor market.

(AI) capabilities will enable automation of some tasks that have long required human labor. Why can (AI) replace some simple human jobs? For example, advances in robotics are expanding machines' abilities to interact with and sharp the physical world. Combined , (AI) and robotics will give rise to smarter machines that can perform more sophisticated functions than ever before and brings more advantages that humans have exercised. This will permit automation of many tasks now performed by human workers and could change the shape of the labor market and human activity.

1.2 How (AI) influences labor market

Today, it may be challenging to predict exactly which jobs will be most immediately affected by (AI)-driven automation. Because (AI) is not a single technology, but rather a collection of technologies that are applied to specific tasks.

Some specific predictions are possible based on the current (AI) technology. For example, driving jobs and house cleaning jobs, bank counter service jobs, telephone enquiry service operators. Restaurant cooking jobs, simple accounting record service jobs etc. that require relatively less education to perform. Advancements in computer vision and related technologies have made the feasibility of fully appear more likely, potentially displacing some workers in driving-dominant professions. Seemingly similar robot, for which the operational tasks is less specific of navigating to a specific destination when following a set of given rules and preserving safety.

In the future, the effects of (AI) on the labor market in the decade ahead will continue the trend toward skill-biased change that computerization and communication innovations have driven in recent decades. Thus, some human driving occupation will be disappeared or replaced by (AI) automation driven. For example, bus drivers, light truck or delivery services drivers, heavy and tractor-trailer truck drivers, school drivers, tax drivers, travel bus drivers.

However, (AI) technology could enable some workers to focus time on other job responsibilities, boosting their productivity, and actually raised wage growth among those still holding the reshaped jobs. For example, salespeople, who currently spend a considerable amount of time driving could find themselves able to do other work when a car drives them from place to place, or inspectors and appraisers could fill out paperwork, when their car drives itself. This (AI) -driven technology should make these workers more productive, with (AI) -driven technology serving as a complement, not a substitute. New jobs will also likely be created, both in existing occupations cheaper transportation costs with lower prices and increase demand for products and all the related occupations, such as service and fulfillment, and in new occupations not currently foreseeable.

What kind of jobs will be created by (AI) technology? Predicting future job growth is extremely difficult, due to it depends on technologies or substitute for existing today as well as they may complement or substitute for existing human skills and jobs. However, (AI) will also lead to substantial indirect job creation to the degree it raises productivity and wages, it may also lead to higher consumption that would support additional jobs from high-end draft production to restaurant and retail. The future(AI) " augmented intelligence", the technology's role is as assisting and expanding the productivity of individuals rather than replacing human work. Thus, based on the biased-technical change framework, demand for labor will likely increase the most in the areas where humans complement (AI) automation technologies. For example, (AI) technology , such as IBM's Watson may improve early detection of some cancers or other illnesses, but a human healthcare professional is needed to work with patients to understand and translate patients' symptoms, inform patients of treatment options, and guide patients through treatment plans. Shipping companies may also partner workers who pick up and deliver products over the last feet with (AI) enabled autonomous vehicles that move workers efficiently from site to site. In such cases, (AI) augments what a human is able to do and allows individuals to either be move effective in their specially task or to operate on a larger scale. Thus, it seems (AI) technology will also create new jobs, raise productivities and workers' efficiencies.

Redefining management in the workforce of artificial intelligence

2.1 Change management

In the future, due to artificial intelligence influences to some kind of human jobs nature. So, the kind of human jobs of management methods will also need to change to adapt the artificial intelligence technology input to their organizations. It will cause challenges for every executive and manager if who won't have effort to manage their teams how to apply artificial intelligence technology to work efficiently and easily. For example, division of labor will change among humans and machines will increase. Thus, companies will have to adapt their training performance and talent strategies how to emphasize on work that how to make human judgment and skills and experimentation. Thus, (IA)'s greatest impact will be on administrative coordination and control tasks, such as scheduling , resource allocation.

In fact, mangers will encounter this challenges: How to apply human experience and expertise to judge critical business decisions and practices when the information available is insufficient to suggest a successful course of action? Due to this kind of work will require new skills and mindsets. I shall indicate these change management methods to adapt (AI) technology. Such as: administration and routine tasks, scheduling , allocation of resources and reporting will fall within the intelligence machines, responsibilities that have long been reserved for humans. For example, a typical store manager or a lead nurse at a nursing home most constantly arrange shift schedules, accounting for staff members' absences owing to illness, vacation time or sudden departures.

Thus, the managers need to learn how to arrange new division of labor within the organizations after (AI) technology had been implemented to the organization. Artificial intelligence is currently influencing into once considered exclusive to humans: assessing and acting on human emotions and personality traits. The influences to managers need to change their strategies to adapt (AI) technology implements include such as below:

Firstly, managers need to spend the bulk of their time on coordination and control tasks from intelligent system implements. Their time spending on these major three aspects from impact of intelligent system: coordinate and control, solve problems and collaborate and people and community , strategy and innovation three aspects. Thus (AI) will influence managers need to change their judgment method to teach whose teams how to adapt the (AI) system operations in any organizations.

Secondly, (AI) will influence top, middle and low level management needs to change to adapt the (AI) technology operations to any owned (AI) technology organizations in the future. Intelligent machines must be trained in context. Just like humans , on-the-job training is a requirement for such machines because they typically arrive with only very general capabilities. To get the most from (AI), managers at all levels must participate in the instructional experience and in the learning process and provides managers' familiarity with such systems on these aspects, e.g. How the system works and generate advice, how the system has a proven track record , how the system provides convincing explanations , how the system can make simple rule- based decisions.

Thirdly, managers need to learn how to make judgment more accurate (AI) systems assistance. Although (AI) will invariably take on more routine work and even augment human decision-making, it won't judgment work, the application of human experience and expertise to critical business decisions when the information available is insufficient to suggest a successful course of action or reliable enough to suggest an obvious course of action. For a sense of the nature of judgment work, consider big data marketing and sales analytics. Such analytics often provide insights that can inform promotional campaigns, including predicting which promotions will generate desired sales brand further into the future, marketing executives need use judgment, combining analytics with their own and others' insight and experience.

The application of experience and expertise to critical business decisions and practice represents the real value of human judgment. But, when artificial intelligent machines are implemented to any organizations to assist the low, middle and top level management to make any business judgment. These forms of judgment work that managers can gather data interpretation, idea development more absolute from (AI) machine assistance. Thus, why these level management executives need to learn how to apply (AI) machines to help them to make any business judgment more accurate.

2.2 How (AI) influences organizational change

Consequently creative and social intelligence will be in even greater demand as (AI) makes in management and the workforce. This development will represent a long term trend in labor markets , one characterized by intensifying demand and reward for social skills with a growing desire for creative capabilities, managers will seek to fashion of ideas and hypotheses from inside and outside of the enterprise to shape solutions to their most pressing business problems. Thus, (AI) will influence overall organizational team members who have chance to participate any decision to make more accurate business judgment.

Many managers mistakenly view judgment work as only an individual discipline, failing to appreciate that it can also involve decide interpersonal and organizational practices. In more complex settings, judgment is typically a collective outcome of individuals' and teams' diverse perspectives, insights and experiences. And often , the resulting choices are better informed than decisions that an individual would have arrived at on his or her own.

Thus, when any organizations apply (AI) technology to assist managers to gather data and ideas to make any judgment. In these cases, organizations can create the conditions for effective collective judgment by establishing structures , such as " shadow advisory boards" that prompt managers and employees to source and synthesize multiple perspectives. Thus, a traditional organization (firm) might freshen its thinking is t put together a shadow advisory board, comprised of young, digital people who can apply (AI) machine assistance to make judgment work more accurate whether related to people development, problem-solving or strategizing and innovating for considerable degrees of creative and social intelligence.

Thus, on the one hand, (AI) technology machine augmentation and automation can give these advantages to human (organization managers) , e.g. developing people and community, solving problems and collaborating, coordinating and controlling work, shaping strategy and leading innovation. Besides, on the other hand, the next generation managers need have these individual attitude to treat intelligent machines to be as colleagues.

When, judgment is a human skill, intelligent machines can accelerate human learning that supports it, assisting in data -driven simulations, scenarios and search and discovery activities. Focuses on judgment work, some decisions require insight beyond what data can tell them. This is the sweet sport for human judgment, the application of experience and expertise to critical business decisions and practices. Thus, managers will also need to find ways to learn how to use digital (AI) technologies to tap into the knowledge and judgment of partners, customer external stakeholders and role models in other industries after the (AI) machine had been implemented to the organization.

Future works change:Automation, employment
and productivity

3.1 How (AI) influences employment

Human future " micro to macro" industry trends will be affected business strategy and public policy by (AI) technology. In the future (AI) technology will influence those six themes: productivity and growth, natural resources, labor markets, the evolution of global financial markets, the economic impact of technology and innovation and urbanization. However, (AI) technology will bring economic benefits of tackling gender inequality, a new global competition, Chinese innovation and digital globalization.

Nowadays, advances in robotics artificial intelligence, and machine learning are in a new age of automation, as machines match or outperform human performance in a development to any countries. For example, automation of activities can enable businesses to improve performance by reducing errors and improving quality and speed, and in some cases achieving outcomes that go beyond human capabilities. For example, some research indicated automation could raise productivity growth globally by 0.8 to 1.4 % annually; more than 2,000 work activities across 800 occupations. When less than 5% of all occupations can be automated using demonstrated technologies about 60% of all occupations have at least 30% of constituent activities that could be automated. Many occupations will change that will be automated away: Activities most susceptible to automation involve physical activities, in highly structured and predictable environments, as well as the collection and processing of data. They are most prevalent in manufacturing , accommodation and food service and retail trade and include some middle-skill jobs. For example, such as natural language processing is a key factor. Beyond technical feasibility, the cost of technology competition with labor including skills and supply and demand dynamics, performance benefits including and beyond labor cost savings, and social and regulatory acceptance will be affected by (AI) automation technology. Thus, (AI) automation will impact to influence global employment in those aspects as below:

Firstly, assuming that people are displaced by automation will find other employment. The anticipated shift in the activities in the labor force is of a similar order as the long-term shift away from agriculture and decreases in manufacturing share of employment. Both of manufacturing and agriculture industries which would be accompanied by the creation of new types of work not foreseen at the time.

Secondly, for business, the performance benefits of automation are relatively clear. Thus, the businessmen have opportunities for their micro economies to benefits from the productivity growth potential and macro economies to benefit to encourage continued progress and innovation , investment and market incentives. At the same time, employers must innovate policies to help workers and institutions adapt to the impact on employment.

This will likely include rethinking education and training, income support and safety nets , as well as support for those dislocated, when employees need to leave themselves homes to move to other cities to learn new (AI) automation works. Thus, individuals in the workplace will need to engage move comprehensively with machines as part of their everyday activities, and acquire new skills that will be in demand in the new automation age. Consequently , the scale of shifts in the labor force over many decades that automation technologies can be a similar order to the long -term technology -enables shifts in the developed countries' workforces away from agriculture in the 21 th century. Those shifts did not result in long-term mass unemployment because they were accompanied by the creation of new types of work not foreseen at the time. However, human will still be needed in the workforce when the total productivity gains are caused by (AI) technology.

3.2 What occupations will be influenced by (AI) technology.

In the future, scientists predict that these occupations will be influenced by (AI) technology mostly. They include : retail salespeople, food and beverage service workers, language or translation teachers, health practitioners. Since these work activities have a more relevant occupations are made up of a range of activities with different potential for (AI) automation . For example, a retail salesperson will spend more time interacting with customers, stocking shelves , or ringing up sales. Each of these activities is distinct and requires different capabilities to perform successfully.

Thus, these job activities have similar simple control characteristics. Simple activities include greet customers, answer questions about products and services, clean and maintain work areas, demonstrate product feature process sales and transactions. All these activities can have similar simple activities in order to (AI) machines can be learn how to do these activities from (AI) technology . For example, the capability perception includes sensory perception, cognitive capabilities, such as retrieving automation, recognizing known patterns(supervised learning), logical reasoning problem solving.

Thus, (AI) machine is such human, which has feeling and emotion, such as social and emotional sensing, judgement reasoning methods, natural language understanding and physical capabilities, such as mobility , navigation, gross motor skill, fine motor skills. It seems that the future, (AI) human invents machines which will have these human characteristics to do human similar behavioral job duties more easily and efficiently. It implies these above human occupations will be replaced by (AI) human invention machines in the future. Due to (AI) creation, it is possible to cause unemployment number of these above workers will increase because (AI) machines can do their similar job behavioral activities.

Consequently, employers won't need to employ many of these skillful labor. Otherwise, they can buy less number (AI) machines to attempt to do whose job activities more easily and efficiently. So, it seems (AI) machines will have more high work performance to replace these occupation workers' work performance. Finally, these occupation worker unemployment number will only increase when the (AI) machines had been invented to achieve to do their work behavioral activities absolutely success in the future.

3.3 Whether (A) technology machine labor will replace human worker more or assist human worker more

There is no single agreed definition of a robot how outcome of a task that is completed without human intervention. When some definitions require the task to be completed by a physical machine moves and respond to its environment, other definitions use the term robot in connection with tasks completed by software , without physical embodiment.

However, to answer the question : Whether (AI) technology machine labor will replace human worker more or assist human worker more. I shall indicate some examples to let readers to judge whether (AI) technology can create new jobs or reduce old jobs.

Firstly, I shall explain what (AI) function is. (AI) is a service robot that performs useful tasks for humans or equipment excluding industrial automation application . Thus, the classification of a robot into industrial robot or service robot is done according to its intended application. It is also a personal service robot or a service robot for personal used for a non commercial task, usually by lay persons . Examples are domestic servant robot, and pet exercising robot. It is also a professional service robot or a service robot for professional used for a commercial task, usually operated by a properly trained operator. Examples, are cleaning robot for public places, delivery robot in offices or hospitals, fire-fighting robot, rehabilitation robot and surgery robot in hospitals. Thus, these functions will be future (AI) application to our daily life necessaries or business necessaries.

However, some authors agree (AI) will bring negative outcomes of automation, due to raise competiveness, reduce human job nature. Otherwise, other authors argue (AI) will bring positive outcomes of automation, due to raise productivities, job creation, assist humans work.

On the positive outcome hand, robots can increase productivity . This is particularly important for small-to medium sized businesses both are in developed and developing countries economies. It also enables large companies to increase their competitiveness through faster product development and delivery. Increased use of robot is also enabling companies in high cost countries to re shore, or bring back to their domestic base parts of the supply chain that will have previously outsourced to sources of cheaper labor. Currently , the greater threat to employment is not a automation, but an inability to remain competitive. Automation has led overall to an increase in labor demand and positive impact on wages. The reason is that the middle-income/middle-skilled jobs have reduced as a proportion of overall contribution to employment and earnings leading to fears of increasing income inequality, the skills range within the middle income bracket is large. Thus, robots are driving an increase in demand for workers at the higher -skilled and with a positive impact on wages. This issue is how to enable middle-income earners in the lower-income range to unskilled or retain. Finally, the (AI) positive impact supporter who argue the future will be robots and humans can work together.

However, on the negative outcome hand, robots can substitute labor activities, but don't replace jobs. They believe that less than 10% of jobs are fully automatable. Increasingly , robots are used to complement and augment labor activities, the net impact on jobs and the quality of work is positive. Automation can provide the opportunity for humans to focus on higher-skilled, higher-quality and higher-paid tasks. Robots can improve productivity when they are applied to tasks that which perform more efficiently and to a higher and more consistent level of quality than humans. For example, increased productivity is enabling some firms, such as Whirlpool, Caterpillar and Ford Motors company in the US restructure their supply chains, bringing back parts of the manufacturing process to the country of origin. Thus, productivity gains due to robotics and automation are important not just at the company level, but also for build industry and nation competitiveness.

I suppose that productivity can be raised. What are the impacts of robots on employment? Firstly, the main focus of development has been on personal entertainment, which does not drive worker productivity (manufacturing production). When the internet (information and communication technology (ICT)) innovation. This is borne and by findings that manufacturing productivity, which has been driven by innovations in automation rather than consumer technologies, has government strongly than productivity in the services sectors of the economy in most nature economies. It seems (AI) automation will create many jobs in internet communication entertainment game industry. For example, many young people like to use internet to play any electronic games from computer or mobile at home or outside home conveniently. Thus, (AI) automation will increase demand to be invented to any new entertainment game from internet channel. It will need to employ many (AI) entertainment game inventors to create many automation entertainment games. Thus, (AI) automation in internet entertainment game industry will need human (AI) entertainment game inventors to invent the knowledge-based capital of (AI) automation entertainment games. The (AI) entertainment game inventors will need own research and development skills, form specific skills, organizational know-how skills, databased knowledge, design and various forms of intellectual property to do these (AI) automation entertainment game invention occupations in the future.

International Federation Of Robotics(2016) indicated that China will be as a major robotics manufacturer and user of robots, benefiting from jobs created by robot manufacturing and productivity gains from robot use. Chins had sold of robots to any one single market every year since 2017 year. The Chinese government has included a focus on robotics in its 10 year strategy. In order to achieve its target of a robot density of 150 units per 10, 000 workers by 2020 year. Thus, Chinese companies will have to install around 650,000 new industrial robots between 2016 to 2020 year, 2.5

times more than installed globally in 2015 year.

Hence, China (AI) manufacturing industry will need to employ many workers . It implies (AI) manufacturing industry will create many new occupations in China. Also, ministry of economy, trade and industry (2015) also showed that Japan currently has the largest stock of industrial robots in operations, primarily in the automation industry. Driven by a rapidly aging population and low productivity rates, the Japanese government has sights on a 20-fold increase in the use of robots in the non-manufacturing sector and a three-fold growth rate of labor productivity in the service sector both by 2020 year. Thus, it also implies Japan will need many robots to be provide to service industry. Due to robots will provide to serve any businessmen's clients. Thus, it is possible that the service workers won't be dismissed as well as it is depended on the serving job nature to decide whether Japan's service workers can still serve to their employer when the service (AI) robots are applied to whose employers.

Consequently, it seems that (AI) can create employment, Ministry of economy, trade and industry (2015) showed that such as China will develop the major (AI) automation manufacturing industry. The (AI) employers will need to employ many workers to manufacture any these different kinds of (AI) robots to satisfy China or overseas individual or business buyers needs. But, (AI) can also cause unemployment to the low skillful service workers. Such as if Japan some service businesses choose to buy any (AI) service robots to replace their service staffs to serve their clients. It is possible that the service staffs will be dismissed, due to (AI) robots can do such as their same service job duties to achieve better service performance.

Thus, today, it is increasingly common for people to use robots in various situations at home and in retail stores, hotels and hospitals these service industries. Robots are classified into server types based on their functionality (service and utility robots or those designed to communicate with humans) and appearance (humanoid robots or mechanical robots). The type of robot, to which each country allocated particular importance in the advance of robotics, reflects the sense of values and preferences of its population. Thus, if the country has high population needs to use robots, then they will influence either more new jobs creation or more old job loss in the country's (AI) manufacturing or (AI) service industries both. For example, Japan respondents often associate the term " robot " with humanoid robots that can communicate with human and they have a high level of familiarity with robot. The US has the highest level of robot utilization at home and in retail stores with its people being the most enthusiastic about the future use of robots. Germany shows a strong tendency to consider robots for industrial purposes and its people feel strong effort to the presence of robots in their households.

In conclusion, to judge whether how (AI) will influence the country's employment to be better or worse. It will depend on the country home buyers (users) or business buyers (users) how to use (AI) for their daily needs. If the country , such as US retail stores need to use (AI) , it will have possible to reduce some or many retail service workers. Even, if the country , such as Japan has many home users need to use (AI) , it will not influence the employment market. Otherwise, it will raise (AI) salespeople numbers. Even, if the country, such as Germany and China will have many (AI) manufacturers, then it will create many (AI) manufacturing occupations for these (AI) manufactory workers.

Consequently, (AI) robots manufacturing and service needs will have positive or negative impact to any country's employment. It will depend on the (AI) service provision and service workers' job nature as well as the manufacturing workers of (AI) knowledge level to decide their employment chance in their country's employment market.

What does artificial intelligence(AI) mean?

- What (AI) function is?

Some scientists explain that artificial intelligence means which is an expert system, computer software that embodies a portion of the specialized knowledge of a human portion in a specific, narrow domain, owns decision making ability of human expert. The (AI)technology is based on the premise that what makes a person an expert is years of experience that enables who recognizes certain patterns in a problem as being similar to pattern. For example, in the future artificial intelligence system can be applied to control air traffic, design to computer configuration, medical diagnosis, instruction/training, speech/interpretation, monitoring to (nuclear plant), planning to mission, factory scheduling, prediction weather, repairing telephone, automatic driving etc. different industries.

Artificial intelligence characteristics include: creative, adaptive , common sense, fact processing, quick replication, broad focus permanent and consistent skill. Otherwise, traditional computer expert system disadvantage includes perishable, unpredictable, slow reproduction, expensive, slow reproduction, slow processing lacks inspiration, needs instruction, narrow focus only machine knowledge. So, artificial intelligence is a branch of computer science devoted to creating computer to influence software and hardware to attempt to create human intelligence or human intelligent behavior. It is learning from experience, responds flexibility in situation that are, new or not anticipated.

Thus, (AI) can be learnt programmed knowledge to solve problems, using reasoning in solving problem, understanding and inferring facts and rules, recognizing the relative importance of different elements in a situation. In summary, artificial intelligence is concerned with two basic ideas mainly: The first idea, it involves studying the thought processes of humans to understand what intelligence is; the second idea, it deals with representing thought processes using companies to create artificially intelligent entities for testing the theories of intelligence.

- Can (AI) impact human job nature?

Human need concern this question: Will artificial intelligence (AI) reduce some human jobs in order to instead of replacing machines to do? Due to artificial intelligence is the ability of machines to do thing, that people would require intelligence. For example, artificial intelligence machine man driving(self-driver), it (AI) machine man driving research is an attempt to discover and describe aspects of human intelligence that can be simulated by driving machine functions. Alternatively, (AI) mathematical research may be another viewed as an attempt to develop a mathematical theory function to describe the abilities and actions of things (natural or man-made) exhibiting intelligent behavior and server as a design of intelligent calculation machine function.

Why do humans need artificial intelligence machines to instead of traditional human service job? For example, can artificial intelligence machine man (self-driving) driver drive to replace human driver? I shall compare the differences between humans and computers : The characteristics of humans are good at recognizing various things, either seen before or not, recognizing the relationship patterns between things. Human thinking is common sense reasoning, combining all types of sensory input, acting appropriately in novel situations, learning new things and changing behavior patterns, making decisions , even when given incomplete information, working with noisy, incomplete information gathering behaviors . However, characteristics of computers are good at: The tasks humans do naturally are extremely difficult for a computer program as intelligent, which must be able to do the same kind of tack as humans do naturally.

Hence, (AI) is an combination of many different success and technologies: Linguistics - computational and socio, philosophy-logic, philosophy of mind and of language, electronical engineering -image and speech processing, pattern recognition, robotics, machine learning, neural networks, optimization scheduling, management information system and decision making. So, it is possible that (AI) can impact human job nature to instead of human working behavior in the future.

How can (AI) influence labor market?

- How can human society job nature

to be changed to artificial intelligent society?

From the first intelligent perspective reason view point, artificial intelligence is making machines " intelligent" acting as humans expect people to act. Artificial intelligence has ability to distinguish computer responses from human responses, it owns knowledge to solve expert problem. From another research perspective reason view point, artificial intelligence is the study of how to make computers do things which, at the moment, people do better (Rich & Knight, 1991, p.3).

(AI) researchers are native in a variety of domains, e.g. formal tasks (mathematics, games), tasks (perception, robotics, natural language, common sense reasoning), expert tasks (financial analysis, medical diagnostics, engineering, scientific analysis and other areas).

From the second business perspective reason view point, (AI) is a set of many powerful tools, and methodologies for using those tools to solve business problems. From a programming perspective reason view point, (AI) includes the study of symbolic programming problem solving and search .

From the third human technological perspective reason view point, today's computer can do many well-defined tasks, for example, arithmetic operations, are much faster and more accurate than human beings. However, the computers' interaction with their environment is not very sophisticated yet. How can human test whether a computer has reached the general intelligence level of a human being? Can a computer convince a human interrogator that it is a human? But before thinking of such advanced kinds of machines, human will start developing our own extremely simple " intelligent" machines.

So, it is possible that human society job nature will to be changed to artificial intelligent society when (AI) technology is developed to the mature stage in the future.

- Why does human need artificial intelligence machines?

One of major division in (AI) is between humans who think (AI) is the only serious way of finding out how we (human) work and human who want companies to do very smart things, independently of how we (human) work. This is the important distinction between cognitive scientists vs engineers. One of another major division in (AI) is between symbolic (AI), which represents information through symbols and their relationships. Specific Algorithms are used to process these symbols to solve problems or deduce new knowledge and connectionist. So (AI) , which represents information in network. Biological processes underlying learning, task performance and problem solving are imitated from human mind behaviors.

Thus, it is possible that artificial intelligence machines can do the better judgicious behavior to compare human.

- How does artificial intelligence influence future working changing in automation employment and productivity aspects?

In the automation changing influence aspect, as companies increasingly use robots on production lines or algorithms to optimize their logistics manage inventory, any carry out other core business functions. Technological advances are creating a new automation age in which ever-smarter and more flexible machines will be deployed on an ever larger scale in the marketplace. However, researching artificial intelligence with how influences human working nature. We need to answer these questions: How will automation transform the workplace? What will the implications for employment? And what is likely to be its impact both on productivity in the global economy and on employment?

Advances in robotics, artificial intelligence, and machine learning are growing in a new age of automation as machines match or outperform human performance in a range of work activities, including ones requiring cognitive capabilities. What factors are determined the changing in workplace adoption by artificial intelligence innovation? What advantages are automation? Automation of activities can be enabled businesses to improve performance by reducing errors and improving quality and speed, and achieving outcomes that go beyond human capabilities.

Some scientists indicated based on their scenario modeling. They estimated automation could raise producing growth globally by 0.8 to 1.4 percent annually. Almost, the activities people are paid almost $16 trillion in wages to do in global economy have the potential to be automated by adopting currently demonstrated technology. According to their analysis of more than 2,000 work activities across 800 occupations. When less than 5% of all occupations have of least 30% of activities that could be automated. They also indicated that technical economic and social factors will determine automation. Continued technical progress, for example, in areas such as natural language processing is a key factor beyond technical feasibility , the cost of technology, competition with labor including skills, and supply and demand dynamics, performance benefits including and beyond labor cost savings and social and regulatory acceptance will affect (alter) the scope of automation.

Other some scientists also indicate U.S. country for example, the anticipate shift in the activities in labor force of a similar order of magnitude as the long term sight away from agriculture and decreases in manufacturing. Share of employment in the United States both which were achieved. So, those factors can influence why artificial intelligence technology needs. So, it is possible that future agriculture and manufacturing both industries will apply

(AI) technology manufacturer-kind of job nature to raise productivity instead of farmers, fruit picking workers, farming transportation labours as well as factory manufacturing workers and supervisors etc. human-kind of job nature.

- Is artificial intelligence possible to replace labor ?

Not just intelligence, but also debating, if machines are capable of having a conscious minds. Artificial intelligence has those characteristics as below:

On functionalism aspect, artificial intelligence inputs mental states, sensory inputs, (beliefs, desires being in pain feeling) and behavioral outputs. Since mental states are identified by a functional role, which are thoughts to be manifested in various systems. Even, perhaps computers which are physical devices with electronic substrate that inform computations on inputs to give outputs similar to brains which are artificial intelligence composed of part any intrinsic relationship to each other. Thus, artificial intelligence activities is not the whole itself, but into parts or on external influence on the parts.

On dualism aspect, artificial intelligence is a set of views about the relationship between mind are matter. On materialism aspect, it builds the only thing that exists is matter, including consciousness.

On biological naturalism aspect, it is similar a human brain than feels pains makes mental situation. So, artificial intelligence is similar biologist which might to be excited to human labor work.

Hence, it seems artificial intelligence can change (alter) or replace human labor work of nature in possible in the future.

- Can (AI) technology replace human labour nature of work?

On technological innovation reason view point, the history development of artificial intelligence studying the intelligence is one of most ancient scientific discipline. The history development of artificial intelligence what aims to achieve human use to sense, learn remember and think, logic probability, decision making and calculation develop from mathematics, instead of replacement human labor functions.

Artificial intelligence history development aim is the scientific analysis of skills in connection and practice with the appearance of computers from 1950 year beginning. The artificial intelligence (AI) can deal with the ultimate challenges. How can (either biological or electronic) mind sense, understand and manipulate a world that is much simple and more complex than itself? And what if would human like to construct something with such capabilities?

The general-purpose software of the early period of (AI) were only able to solve simple tasks effectively and failed when which should be used in a wider range or an more difficult tasks. One of the sources of difficulty was that early software had very few or mix knowledge about the problems which handled, and activities successes by simply syntactic manipulation. Moreover, the other difficulty was that many problems that were tried to solve by the (AI) were untreatable.

The early (AI) software whether trying step sequences based on the basic facts about the problem that should be solved, experimented with different combinations till which found a solution. From the end the 1960 year, developing the so-called expert systems were emphasized. These systems had (sue-based) knowledge base about the field which handled. Till to the beginning of the 1970 year, (Prolog) the logical programming language was born, which was built in the computation realization of a version of the resolution calculus. (Prolog) is a remarkably prevalent tool in developing expert systems (on medical, judiciary and other scopes), but natural language parsers were implemented in this language. Then, in 1981 s, the Japanese announced the fifth generation computer system project, a 10 years plan to build an intelligent computer system that use the (Prolog) language as a machine code. Nowadays, (AI) can be applied any industries, such as car manufacturing industry can use (AI) technological machine-men manufacture car, instead of replacing human labors in factory. Even, in the future, using (AI) machine-men drivers can drive any private cars or public transportation tools, instead of replacing human drivers, e.g. bus, train, tram, ferry etc. Also in the future, machine-men can replace housewives to serve families to do housekeeping clean job , e.g. cleaning toilets, bathrooms, kitchens, even cooking functions at home. So (AI) machine-man can reduce housewives works at home. Moreover, (AI) machine man can take care old people , when who are living at homes or elder care centers.

So, it seems artificial intelligence (AI) will be possible developed to manufacture a new generation machine-man to assist (serve) families to do any simply cleaning or cooking jobs at homes. Moreover, the overall demand of (AI) general social needs will also rise, such as security, driving transportation tools, restaurant cleaning, elder centers care service etc. So, it seems that individual or families or social needs of (AI) will be increase in the future. Thus, it will influence macro economy growth (GDP) if there are large house family consumer group and hotel or bus or taxis or ferry etc. different business consumer group demand any artificial intelligence machine numbers increasing. Then, the artificial intelligence products and material manufacturers must need to buy many artificaial intelligence materials to produce any kinds of artificial intelligence machines to prepare to satisfy consumer individual needs. Consequently, macro economy will grow to the owned artificial intelligence development countries, e.g. US, China, UK.

● Why can artificial intelligence satisfy human needs?

First, On machine-man satisfactory demand aspect view point, it makes computers that think, it is the automation of activities. We associate with human thinking: like decision making, learning. It is the act of creating machine that perform function that require intelligence when performed by people. It is the study of mental faculties through the use of computational models. It is the study of computations that make it possible to perceive, reason and act. It is a branch of computer science that is concerned with the automation of intelligent behavior. It is anything in computing service that human don't yet know how to do property.

Second, on thought aspect artificial intelligence means systems thank think like humans, systems that think rationally.

Third, on behavioral aspect, artificial intelligence systems that act like human and that systems act rationally. However, the basic objective of (AI) is to represent human's thought processes in computation . These machines are supposed to exhibit behavior that. It is performed by a human being, would be considered intelligent. However, some authors feel (AI) has disadvantages, such as it is not creative, it is excited in the use of sensory devices, it can't make use of a very wide context of experiences and it does not use common sense.

For speech recognition and understanding function needs example, (AI) can be applied in speech recognition and understanding function, which (AI) speech or voice recognition is a data input method. For example, the computer recognizes and understands one (or a few) word commands. Speech understanding on the other hand is the computer's ability to understanding a spoken language. That is , the computer understands the meaning of sentences, an paragraphs through (AI).

So, (AI) can be attempted to learn human language how to speak. It is similar to translate human language skill, instead of actual human speaking skill. Also, (AI) can assist handicap learning or language student how to listen different languages by machine-man sounds from computers more accurately.

So, it seems that it (AI) can replace human language teachers speaking function and can change teaching language nature of job in language speaking and listening education industry.

- Is artificial intelligence one good choice for human future technological benefit?

Nowadays, new technology development is popular. However, artificial intelligence is one kind of new technology choice among different technologies innovation. So it brings this question: Is artificial intelligence technology value to invest? To answer this question. I shall indicate some other new technology developments to compare (AI) technology development to judge which has urgent needs to achieve human expectation nowadays.

For example, why is green peace interested in new technologies? New technologies features prominently in our ongoing campaigns against genetic modified crops and number power. However, which are also an integral part of our solutions to environmental challenges, including renewable energy technologies, such as solar, wind and wave (water) power energy as well as waste treatment technologies, such as mechanical, biological treatment.

It seems humans need concern how to apply (AI) technology to solve environment pollution challenges in our future. So, environment protective, agriculture, natural energy technology will be popular demand to attempt to apply (AI) technology to solve their challenges or apply (AI) to assist to develop their industry.

What is relationship between
(AI) and economy growth?

- How can artificial intelligence technology influence economy?

Advances in artificial intelligence (AI) technology and related fields have opened up new markets and new opportunities progress in critical areas, such as health, education, energy, economic development, social welfare and the environment pollution.

(AI) automation will continue to create wealth and expand the global economy development in the future. However, when many will benefits that growth won't be costless and will be accompanied by changes in the skills, that workers need to increase productivity in the economy and structural changes in the economy. So, in the skills that workers need to succeed in the economy and structural changes.

I shall indicate why aggressive policy action will be needed to help Americans who are disadvantaged by these changes , due to (AI) technology is caused. For automation industry change example, artificial intelligence (AI) capabilities will enable automation of some tasks that have long required human labor. These artificial intelligence technology introduction can increase new opportunities for individuals. The economy and society, but (AI) has also the potential to disrupt be current livelihoods of many Americans. However, (AI) leads to unemployment and increase in inequality over the long run depends not only on the (AI) technology itself, but also on the institutions and policies that are changed. Thus, it is possible that (AI) technology will raise some countries unemployment number if the employer apply (AI) technology workers to work instead of human labor in their factories, but it can also raise productivities for these employers.

- Can (AI) influence global economy growth?

Technological progress is main driver of growth of GDP per capita, allowing output to increase faster than labor and capital . However, technology can increase productivity, but also decrease the number of labor hours needed to create a unit of output. So (AI) causes unequal to labor wage decreases, even reduces the number of labor to manufacture, e.g. artificial intelligence technology of automation car manufacturing industry; clothing manufacturing industry; plane manufacturing etc. high technology of artificial intelligence manufacturing method. But (AI) should be potential environment benefit, although it raises unemployment ratio. Moreover, it can rise production , due to many skilled craft were replaced by the combination of machines and lower-skilled labor. The result of (AI) technology introduction , it causes output per hour risen when inequality declined, driving up average living standards, but the labor of some high-skill workers was no longer as valuable in the market. Otherwise, if (AI) technology is continue developed to be success. Some routine intensive occupations will be loss, which focused on predictable, e.g. easily programmable tasks, such as switchboard operators, filing clerks, travel agents, and assembly line workers would be particularly replaced by new (AI) technology. However, at the same time, (AI) technology development will bring these benefits: improvement in education (training (AI) technology scientists) , due to (AI) manufacturing technology needs are raising to businesses and institutional changes, such as the reduction in unionization and raising in the minimum wage to the (AI) manufacturing technology skilled labor in factories.

Because (AI) technology is not a single technology, but rather a collection of technologies that are applied to specific tasks, the effects of (AI) will be felt unevenly though the economy. It will bring some tasks will be most easily automated than others , and some jobs will be affected more than others, both negatively and positively. Finally, new jobs are likely to be directly created in areas , such as the development and supervision of (AI) as well as indirectly created in a range areas though out the economy as higher incomes lead to expanded demand.

However, if (AI) technology could dominate global labor markets. If labor productivity increases, do not influence into wage increases, then the large economic gains brought about by (AI) technology could be increased wealth inequality, due to employers can reduce production cost, but workers (labors) wages will not be increased, even will be decreased. Hence, it seems the (AI) technology will bring disadvantages to labor market to cause unemployment or reduce wages in possible, although it can reduce employer individual salary (wage) expenditure and it can raise productivity.

- How can artificial intelligence impact global economy growth?

Artificial intelligence (AI) technology is a branch of computer science that aims to create intelligent machines that work and react like humans. So, (AI) is a technology that appears to impact (influence) human preference by learning, understanding complex contents, enhancing humans in executing both routine and non-routine tasks. In the future, (AI) technology that can be virtual personal assistant, as well as it may exist, such as robots with human-like processing capabilities.

How can (AI) technology impact global job nature and job behavioral changing over the next 10 years? During this time period, (AI) technology is predicted to have wide-ranging applications including: Machine learning that automates analytical model building by using algorithms that allow machines to operate without human assistance. In global education aspect, potential applications include predicting cause-and-effect relationships from biological data, identifying new drugs, self-driving cars, and protecting against fraud, improved natural language processing that allows computers to continue to better analysis, understand and generate language to interface with humans using natural human languages. For example, transcribing notes dictated by physicians, automatically drafting articles and translating text and speech. So (AI) technology can be applied to education aspect to improve humans' knowledge level.

In visual art aspect, (AI) machine vision that allows computers to identify objects, scenes and activities in images. Current applications of (AI) machine vision include providing objective descriptions for the blind seeing(visual) needs.

We except the economic effects of (AI) technology to include both direct GDP growth from sectors that develop or manufacture. (AI) technology and indirect GDP growth through increased productivity in existing sectors that employ some form of (AI). If (AI) technology is an increasingly critical component of more products, it will become an integral part of many people's lives. Thus, (AI)'s ability to influence economic activity, rather than the economic or development status of the region. (AI) has the potential to impact income classes and to bring significant gains to both developed and developing countries. For example, (AI) has the potential to optimize good production around the world by analyzing agricultural regions and identifying what is necessary to improve crop yields.

In estimating the future economic effects by (AI) technology innovation, it is important to note that it is challenging to accurately predict which applications of (AI) will ultimately be commercially successful. In micro level economic influence, we need to apply methodologies to estimate the economic effects of investment in firms developing (AI) technology since investment levels in a technology are a telling sign of the future potential of that (AI) technology.

- How can (AI) influence GDP of high income countries in the next ten years?

How (AI)'s development may affect the global economy over the next ten years. In fact, (AI) technology has the potential to affect business across the global in a wide range of industries in ways only a number of technologies have done in the parts. For example, (AI) technology's expected to be a useful tool for enhancing human capabilities and in some instances replacing functions, such as driving a car, adoption of broadband internet, mobile telephone, industrial robotic automation have served to enhance human capabilities.

However, significant public debate has focused on projections of (AI) technology's effect on the labor force. However, large companies prefer to invest in (AI) technological industry. For example, face book's (AI) research lab., google machine intelligence lab. and micro soft machine learning and artificial intelligence research division are all making advances in (AI) technology and investing in the industry's top talent. Additionally, between 2010 year and 2015 year, nearly $5 billion in venture capital funding invested in firms across the global developing and employing (AI) technology (Facebook (AI) Research).

- How can artificial intelligence impact on workplace?

Modern information technologies and the labor economy growth of machines is powered by artificial intelligence have already strongly influenced the world of work in the 21 ST century. Computers, algorithms and software simplify every tasks and it is impossible to image how most of our life could be managed without them. How can be the information economy characterized by exponential growth replaces the most production industry based on economy of scales? What will the future world of work look like and how long will it take to get? Will the future world of work be a world where humans spend less time earning their livelihood? Alternatively, are mass unemployment, mass poverty and social distortions also possible scenario for the future, where robots, artificial intelligence systems play an increasingly central role? These questions concern how artificial intelligence further development . Can influence labor economy growth on workplace ? When the labor market has widespread impact on intelligence property, information technology, product liability, competition and labor and employment laws.

How (AI) technology impacts on labor workplace.

The future influence any organizations how labor economies use of (AI) can be analyzed, such as deep machine learning is based on a set of model high level data. Unlike human workers, the machines are connected the whole time in workplace. If one machine makes a mistake, all autonomous systems will keep this in mind and will avoid the same mistake the next time.

Over the long run intelligent machines will win against every human expert. Production robots have been replacing employees because of the (AI) technology. They work more precisely than humans and cost loss. Creative solutions like 3D printers and the self learning ability of these production robots will replace human workers, the automatic data recording and data processing, traditional back office activities are no longer in demand. Autonomous software will collect necessary information and will send it to the employee who needs it. Additionally, dematerialization leads to the phenomenon that traditional physical products are becoming software. For example, CD or DVDs are being replaced by streaming services. The replacement of traditional event ticket, e-travel ticket service products or hard cash will be the next step, due to the possibility of payment by smartphone. So, (AI) technology will impact human's daily life consumption behaviors in the future. For another example, transportation tools, such as boats and ferries and private vehicles will use sensors and navigating without human input. Taxi and truck drivers will become obsolete, the stock store applies to stock managers and postal carriers of the delivery is distributed by (AI) machine delivery method.

What is the relationship between (AI) and (CRM)?

- Can (AI) technology impact on customer relationship management (CRM) ?

Nowadays , (AI) is a technology almost as old as the computer industry itself, it is similar with the advent of personal assistants function to businesses and personal promotion channel, such as (Amazon's Alexa, Apple's Siri, Google's

Assistant) image recognition (face book), personalized recommendations (Netflix , Amazon). Those innovations have been driven by a increase in processing power, lower cost hardware, and the exploding creation and availability of data. It seems, (AI) technology can impact global customer service management method.

How to forecast economic impact modeling to (AI) will affect global economy? Can human forecast business revenue growth and job creation (or destruction) based on (AI) applied to customer relationship management (CRM) activities? In addition to the economic impact on (AI) or (CRM) which can include an estimate of the economic impact attributable to sales forces customer base. What can economic benefits be brought to (CRM) from (AI) technology?

Artificial intelligence(AI) comprises a set of technologies that use natural language processing, machine learning, knowledge graphs, and other tools to answer questions, discover insights and provide recommendations. Computer systems can use (AI) hypothesize and formulate possible answers based on available evidence can be trained through the ingestion of vast amounts of content, and automatically adapt and learn from (AI) self mistakes and failures.

So, any business organizations (customer service departments) can provide efficient and effective customer relationship management of excellent customer service quality if which applied (AI) technology system. The different type of (AI) systems include: (AI) system platforms, machine learning (AI) based data preparation and enrichment tools, machine vision/image recognition, voice speech recognition, text analysis and natural language processing, bots , e.g. face book website and virtual digital assistance solutions, social media pattern analysis , sentiment analysis, advanced numerical analysis (e.g. IOT streaming , machine logs), supporting technologies, knowledge base dialog management, Q&A processing etc. different (AI) technology system customer relationship management (CRM) tools.

(AI) (CRM) of activity can include these categories, such as: corporate marketing, marketing operation, field marketing, customer support, digital commerce, customer analytics, customer influenced product or service design, product or service pricing, finance information, presentation, customer billing, inventory , logistics and fulfilment support, partner management etc. different CRM tools.

(AI) technology of CRM has been carrying on plan different stages to achieve CRM personal assistant tool for businesses. The stages are such as, in the beginning stage of (AI) projects in place, implement now, pilot phase next year in the final stage of (AI) customer relationship management tools are foreseeable future. So, this CRM technology has been improved to plan in different stages every year to prepare to achieve full capacity of CRM service quality for businesses to use in the future.

Hence, how to develop an estimate prediction of the economic impact (AI) technologies could have CRM activities, which depends on gathering macroeconomic information on business revenue and the basic marketing of business revenue and the basic markup of business expenses by major functions (customer support, marketing and sales , production etc.)

An economic impact model that can gather data together and forecast the results how (AI) artificial intelligence technology brings (CRM) customer relationship management benefits to businesses, e.g. surveys investigation includes IT spending by sample countries, GDP and population estimates and forecasts, revenue per employee and ratios of IT spend to GDP. Surveys (questionnaire questions) of forecast results are influenced by (AI) impact can include: results are projected from surveys and rely on estimates are made by respondents on the expected financial improvements in categories of (AI) –assisted customer relationship management activities. The forecast assumes that these estimates are correct; financial estimates are based on estimates of "first year" improvement from full (AI) implementation; forecasts are from planning to implement any artificial intelligence of customer relationship management (CRM) projects, the improvement forecast is of categories of activity , e.g. corporate marketing , digital commerce, and customer analytics. They are not estimates of ROI for the (AI) software. They rely on conservative estimates to which each of these entities might affect company revenue, expenses or productivity. They also rely on estimates of the penetration of software in customer relationship management activities . Net new jobs created are based on the ratio of new revenue to jobs required to support that revenue . They can assume that 50% of the net new revenue will support increases in labor and the rest will go for capital and other operating expenses that may replace jobs lost to automation.

In the future, some of the ways in micro economic benefits to any organizations. (AI) technology is expected to impact CRM activities include: Spending up sales cycles, improving lead generation and qualification solving customer support problems faster (raising service quality), helping companies improve brand campaigns and recognition, lowering costs of support calls when increasing resolution rates, lowering the cost of recruiting employees and partners, increasing revenue from optimized product marketing, optimizing price, distribution logistics and preventing loss through fraud detection. So, micro economic benefits view point, it seems that (AI) CRM technology can raise any companies economic benefits for care term.

Artificial intelligence enables machines or the in-build software to behave like human beings which allows these decisions and act. The advent of (AI) is leading , talking, making decisions and act. The advent of (AI) is leading to new technologies advances and transforming the economic and employment opportunities for humans in a positive way. (AI) related technologies can facilitate our live. For example, industrial robotics, robotic medical assistants, smart games, financial forecasting software, big data analysis, algorithms in health and bioinformatics, pilotless cargo places, drone ambulances and general purpose and workplace robots and others. (Disruptors technologies: Advances that will transform life, business and the global economy).

Artificial intelligence also known as computational intelligence is defined as " the human –like intelligence exhibited by machines or software. It is theorized that intelligence of humans can be described and intelligence machines or software can simulate it. These machines software can be reasonable , learn, perceive and process information, like human mind and thus facilitate human life. They can think and act for us. So, artificial intelligence is an interdisciplinary field of study including computer science, neuroscience, psychology, linguistics and philosophy.

However, (AI) research and developments have economically impacted many industries, such as robotics, telecommunications, computer applications , health, finance, heavy manufacturing, transportation, aviation, e-service and e-commerce, military , music and movie, toys and games entertainment etc. industries.

In fact, many ideas, systems and technologies have been developing in the world of (AI) technology. However, which are net called or considered (AI) products, rather which are mentioned with their specific names, such as smart graphics, machine learning, e-commerce etc. (i.e. this is called (AI) effect).

What is relationship between
(AI) and digital job nature causes?

- How can (AI) technology influence digital economy?

Nowadays, (AI) related industrial applications will replace most human power in fields, including call centers, customer services and air cargo transportation. (AI) technologies also help weather forecasting based on repeated rainfall pattern (data) recognition, through robotics (i.e. floor cleaning, moving lawns etc.) transporting people and products with unmanned vehicles, sending space unmanned smart shuttles, developing robotic arms, predicting market values in stock exchanges by internet, making homes safer, helping elderly and disabled using robotic servants etc.
Among the (AI) related technologies , there are a few that significance for the impact on society and especially on digital economy . (AI) is particularly influential in machine learning. Such as robotics, transportation, finance, health and bioinformatics, e-commerce , e-games, big online data gathering and internet-of-things. For example, machine e-learning is based in bioinformatics and robots that can learn new skills for better caregiving in healthcare. What is machine e-learning? Machines can e-learn from e-data gathering, coming up generalizations and making decisions to act in certain ways from internet.
There are important applications , such as e-machine perception, electronic online natural language learning processing, online search engines, online bioinformatics, online brain –computer interface, online game playing, online robot locomotion, online advertising, online computations finances, online health monitoring, online DNA classification and decision making, online in chemistry –cheminformatics . So, online machine learning can positively impact productivity and it can enhance information and analytical system from (AI) online channel.
What is robotics? Robotics is one of the most strongly influenced fields in (AI). For example, heavy manufacturing industries, robots and used and man power is replaced for effectiveness, precision, and accuracy, especially in respective or dangerous tasks, including welding, assembling , picking and placing .
So, robots can acquire new skills or adapt the changing dynamic environment. Also, artificial intelligence can be applied in developing transportation. For example, automated vehicles, driver assistance systems , safety systems, collision avoidance systems and public transportation. Moreover, (AI) technology has proven to produce some of the best tools to predict stock market fluctuations from internet data gathering method. It's predictions are based on ever-evolving predictions algorithms and systems learn new models and make connections between historical data and new data to measure stock market trading more accurate from internet data gathering channel.
In health field, especially in health data processing , analysis, decision making support and medical diagnosis. So, online data can show which patients will need what treatment and what alternative drugs could be used more accurate from (AI) online data gathering method. Bioinformatics is an interdisciplinary field combining statistics,

(AI) online technology can help in discovering data patterns and modeling through the application of machine learning, artificial neural networks and genetic algorithms. For example, further (AI) technology development of human genome project of online data sequences.

Online shopping can be facilitated by virtual assistants developed through (AI) technology and these assistants can offer the best advice. (AI) online purchase coming after every product image recommendations and personalization bring important revenue to shopping online sites, like Amazon . Smart computer graphics and games, artificial intelligence is useful in smarter computer, graphics, scene modeling , scene rendering processes in order to create, for example, effective human –robot interactions , online machine learning, online strategic games techniques etc. online computer related (AI) software.

So, online big data analysis and big data does have a critical need in the world of online intelligence machines and software in our future. In other words, (AI) offers online technology to enable online big data analysis to provide industrial organizations with valuable information for effective decision making in short time. For example, what IBM's Watson achieved: this machine used 200 million of structured and unstructured content with a special technology of hypothesis generation, massive evidence gathering, analysis and scoring from internet channel.

Finally, (AI) online technology another related internet invention (internet of things) (IOT) is the network of machines or objects connected through internet. These connected objects can sense their internal and external environment, communicate with each other, can send critical data and finally can make decisions to act or correct their environment from (AI) online technology. For example, factories can monitor and automatically change production processes, hospitals can monitor and regulate the health conditions of their patients , schools can collect data from facilities and cars can send data to car makers from (AI) online technology.

Partner predicts that (IOT) market will create about trillion amount value by 2020 year. Although machines collect big data from their environment, whether which gain an insight or learn from these online data largely depends on the (AI) online machine learning principals and (AI) online technology. In 2013, Mckinsey estimated that disruptive technologies closely related with potential economic impact in 2025 year between $7.1 to $13.1 trillion amount (automation of knowledge work, advanced robotics, autonomous or near-autonomous vehicles).

What is the relationship between
(AI) and global digital economy development ?

● Could work activities in China be automated
making in the nation with the world's largest automation potential?

Can (AI) technology influence China economy? Could China workers be affected and jobs made up of routine work activities and predictable? Will programmable tasks be particularly impact to China employment market ? When impact on labor market is likely to be gradual at the aggregate level, it can be sudden and dramatic at the level of specific work activities, rending some job obsolete fairly. Overall (AI) technology will raise digital skills when reducing demand for medium incomer inequality for China workers. It seems (AI) technology's effect on productivity could be crucial to China's future economic growth as the population ages are increasing.

In China, some biggest technological companies driving significant investments in research and development. Moreover, China is one of the leading global (AI) technology development county. However, China will need to focus on building its innovation capacity. For example, United States and United Kingdom are currently producing more influential (AI) technological research. However, if China planed to achieve (AI) technology success, it's traditional industries will need to develop technical know-how –to and overcoming implementation costs prepare to develop (AI) . When (AI) technology is introduced into China society, China government needs to raise concerning ethical, legal, technological security etc. business questions. Also, surrounding issues include privacy, discrimination, legal liability and regulation. It aims to encourage overseas investors to choose to invest (AI) technological industry to raise GDP growth and manufacturing industries income growth for long term in China.

If China encouraged overseas (AI) technology investment in its country. It is possible to influence China employment market to be changed. Because (AI) technology will impact to influence China people daily life. Due to (AI) technology is introduced to China society, many rich people will prefer to spend to buy any high (AI) technological products for entertainment or learning or machine man driving etc. daily necessity activities. Then it will raise GDP growth and will raise (AI) manufacturers or related-(AI) technological manufacturers profit. It is beneficial to China because it can become one high knowledgeable and (AI) technological economical society.

But it will bring bad influences to raise unemployment chance for the low skillful labor. In labor economy aspect influence , how (AI) technology can influence China low skillful labor unemployment ratio raising. The raising low skill labor unemployment reason is because China low skillful human labors are argued or are replaced by (AI) technology creating new challenges to introduce to influence China society of simply human manufacturing job nature to be changed to be high (AI) technology manufacturing job nature in any China factories. Moreover, when (AI) technology introduction to China, it will cause other related social challenges in China. The varied (AI) related challenges, including the difficulty of creating safe and reliable hardware for sensing and affecting (transportation and education), the challenges of gaining public trust, a low resource comities and public safety and security, the challenges of overcoming fears or marginalizing humans in China employment and workplace and the risk of diminishing interpersonal trust because the low skillful labors won't believe any China employers will give chance to employ them , due to (AI) technology will replace their skills and man manufacturing of productivity is much less to compare to (AI) technology manufacturing method.

- How does (AI) technology influence

the future of employment change?

Are future nature of jobs changed to computerization from (AI) technology? Where are the probability of computing occupations from (AI) technology influence? What is expected impacts of future computing on labor market from (AI) technology influence? John Maynard Keynes's frequently cited prediction of widespread technological unemployment " du to our discovery of means of economic the use of labor outrunning the pace of which we can find new used of labor" (Keynes, 1933, p.3).

In the future, (AI) technology will impact some nature of occupations to change computing. This chance will also influence some countries' economic change. For example, some factory human labors hand routine manufacturing tasks will be changed to computerization of routine manufacturing tasks by (AI) technological machine men hand manufacturing method. it will cause a structured shift in the labor market, with workers reallocating their labor supply from middle-income manufacturing to low-income service occupations.

Arguably, this is because the manual tasks of service occupations are less computerization, as who require a higher degree of flexibility and physical adaptability. So, (AI) technology will influence the human hand labor skillful occupation nature of task cheaper , such as vehicle manufacturing , ship manufacturing, computer manufacturing, steel manufacturing, television, radio etc. home electronic products of heavy machine industry change. Due to (AI) technology machine man will be proper to be used to manufacturing these electronic products when the (AI) technology innovation can develop to the mature stage. Then, any countries manufacturers will choose to use (AI) technology machine man, instead of human hand production.

Supposing the future prices of computing are fallen, seriously, problem solving skills are becoming relatively productive, explaining the substantial employment growth in manufacturing occupations, involving cognitive tasks where skilled labor has a comparative advantage, as well as the increase education needs for (AI) technology computing of machine man subject study.

Prediction of education needs for (AI) technology student numbers will increase, due to manufacturing industry needs many (AI) technology students in future employment market. Another (AI) technology influence if the future (AI) technological innovation, e.g. machine man manufacturing or machine man service industries will both increase demand, then with more sophistic software technologies will be disrupted labor markets by marketing workers redundant.

For publishing industry, what is striking about the case in paper book publishing industry will be unpopular? Due to the electronic book publishing industry will be popular, e.g. Amazon publish . (AI) technology can influence paper book manufacturing method which is replaced by machine man electronic book manufacturing method as well as it will cause the computerization is no longer confined to routine manufacturing tasks. Due to (AI) machine man manufacturing technology will be proper to be used to manufacture any products in short time efficiently and effectively , e.g. electronic book products. In the future, if it is fact to occur this case, such as (AI) technological machine man manufacturing method will be adopted (applied) to manufacture electronic books or any products in possible. (AI) technology will cause many manufacturing workers are unemployed. It is beneficial to employers, who can reduce to spend much wages expenditure to employ manufacturing workers, but it will cause many manufacturing workers loss jobs and reduce income to support whose families lives. It will cause social challenges, e.g. increasing stealing crimes if the manufacturing workers had not other skills to find other jobs to do easily. So, manufacturers need to concern over technological unemployment which will be hardly future phenomenon if who decided to dismiss all manufacturing workers, due to (AI) technology machine men replace to them.

If (AI) technology can be innovated to produce any kinds of machine man to serve any service or manufacturing industries successfully. Then, it will bring these questions: Can future that workers be influenced to be automation employment and productivity by (AI) technology influence? Does it impact to influence the (AI) technology countries' productivity and growth and natural resources development and labor markets and evolution of global financial markets and economic impact of technology and innovation and urbanization etc. issues? How will automation transform the workplace? What will be the implication for employment? What is likely to be its impact both on productivity in the global economy and on employment?

In fact, automatic of activities can enable businesses to improve performance by reducing errors chance and improving quality and speed, and same cases achieving outcomes that go beyond human capabilities. Some economists indicate (AI) technology would give a needed boost to economic growth and prosperity have of the working age population in many countries. Based on the scenario modeling, they estimate automation could raise productivity growth globally by 0.8 to 1.4 % annually. They also indicated that almost half the activities people are almost $1.6 trillion in wages to do in the global economy have the potential to be automated adapting current demonstrates technology, according to their analysis of more than 2,000 work activities across 800 occupations. When less than 5% of all occupations can be automated entirely using demonstrated technology, about 60% of all occupations have at least 30% of worker made activities, that would be automated. More occupation will change to be automated. They also indicated for business performance benefits of automation are relatively clear, but the issues are more complicated by policy making to attract foreign investors. Beyond technical feasibility, the cost of technology, competition labor will include skills and supply and demand dynamics, performance benefits and beyond labor cost savings and social and regulatory acceptance will affect the automation. Their predictions suggest that half of today work activities could be automated by 2055 year, but this could happen 10 to 20 years earlier or latter depending on the various factors in addition to their wider economic condition.

Some scientists suggest (AI) technology is finally starting to deliver real-life business benefits. Computer power is growing significantly , algorithms are becoming more sophisticated and perhaps most important of all, the world is generating vast quantities of the fuel that powers (AI) technology data billions of gigabytes of it every day. Also, online firms are digital natives, such as Google online search service company is investing on (AI) technology. For new though most of the news if coming from the suppliers of (AI) technologies. And many new users are only in the experimental phase. Few products are on the market or are likely to arrive these soon to drive immediate and widespread adoption. As a result, analysts believe (AI) technology's potential will give true economic benefit in the future. (AI) industry will introduce to suppliers and users to raise economic potential of (AI) technology.

In the future, (AI) technology systems can solve business problems. Some scientists categorized those into five technology systems that are key areas of (AI) technology development: robotics and autonomous vehicles, computer vision language virtual agents and machine learning , which is based on algorithms that learn from data without replying on rules-based programming in order to draw conclusions or direct an action.

Such as computer vision and language includes natural language processing, analytics, speech recognition technology, some are about learning from information, such as about machine learning and others are related to acting on information, such as robotics, autonomous vehicles and virtual agents, which are computer programs that can converse with humans. Machine learning and a subfield called deep learning are artificial intelligence applications.

● Can artificial intelligence impact
global economy growth?

Artificial intelligence (AI) is a term first defined in 1956 year. It is a branch of computer science that aims to create intelligent machines that work and react like humans. In contrast today, 60 years later, (AI) is characterized by a number of applications, including computers playing games against humans and understanding human languages, virtual personal assistants, and robotics which involve computers seeing , hearing and reacting to sensory stimuli. In the future, technologists predict for (AI) technology ranging from (AI) being used as a tool to aid relatively simple processes for robots with human like mental capabilities, who expect (AI) technology can emulate human performance by learning, coming to mind its own conclusions, understanding complex content, engaging in dialog with people, enhancing human cognitive performance or replacing humans in executing both routine and non-routine tasks. In existing industry, (AI) technology is used , such as targeted advertising and virtual used personal assistant as well as the (AI) technology that my exist in the future, such as robots with human vehicle processing capabilities.

The range of (AI) technology's progress in the future will determine the economic impact future of (AI) technology on the global economy with more limited advances and applications (i.e. weak (AI) only) corresponding to more limited economic impacts and more substantial progress, i.e. strong (AI) technology is corresponding to more significant economic impact.

(AI) technology learning that automates analytical model, including predicting cause-and-effect relationship from biological data, identifying new drugs, self-driving cars and protecting against fraud etc. functions. Also (AI) learning can improve natural language processing that allows computers to continue to better analyze, understand and generate language to interface with human using the natural human language, virtual personal assistant, helps users by providing scheduling appointment, reminds organizing personal finance and finding providers of various services, machine vision allows (AI) machine man to identify object, scenes and activities in detect pedestrians and bicyclists.

We expect the economic effects of (AI) technology to include both direct GDP growth from sectors that develop or manufacture (AI) technology and indirect GDP growth through increased productivity in existing sectors that employ some from of (AI) technology. If (AI) producing sectors could grow, then it could lead to increase revenues and employment of (AI) technological professionals within these existing firms as well as the potential creation of entirely new economic activities to any countries' societies productivity improvement in existing sectors could be realized through faster and move efficient processes and decision making as well as increased (AI) technological knowledge and access to information available in societies easily.

In the future, if (AI) technology is an increasingly critical component of more products, it will become an integral part of necessary products of many people's lives. The extent of (AI)'s economy effort is also likely to vary from region to region, thought variation may be more dependent on the predominate economic activity of a region and the (AI) ability can influence economic activity, rather then the economic or developmental status of the regions. (AI) technology can move accessibility and can use source development to do international business between one country and another country.

So (AI) technology has the potential to give benefits to different income chooses and to bring significant gains to both developed and developing countries. For agricultural technology, (AI) has the potential to optimize food production around the world by analyzing agricultural regions and identifying what is necessary to improve crop yield. In total, (AI) technology gives greater economic impact to any countries agricultural regions if which implemented (AI) technology to grow crop , fruit etc. food production in the farms.

Investment in (AI) technology is such as capital investment to any countries' public or private enterprises. So, it will have large economic impact to the future . If the (AI) technology is reasonable invested to the different needs aspect by the public or private enterprises in the country. Then, it will have good economic impact to the country in the future. However, when (AI) technology is likely to affect both the productivity and employment components of economic growth in many sectors. Significant public debate has focused on projections of (AI)'s effect on the labor force. However, for instance, some researchers have argued that the rise of (AI) technology and automation will led to significant unemployment as capital is substituted for the low skillful labor. So, they point to the concern that the increasing sophistication of (AI) technology may balance skilled and semi-skilled workers and the reduce the size of the middle class. However, this is not a new argument, due to (AI) technology negatively affecting the labor force and leading to mass unemployment. Because the (AI) technology is the substitution of machinery for human labor. Although, employment in certain industries, has been reduced in the past due to technological advancement. For long term, the labor market has adapted to the introduction of new technology, giving rise to new jobs in new areas. (AI) technology may also be accomplished without a reduction to total employment in the long-term to some Asia countries, such as Hong Kong and Japan. Because Hong Kong and Japan many low skilled labor, e.g. security, cleaner who complaint that employers need them to work long time hours. (abnormal working hours) e.g. one day 12 to 15 working hour per day. Hence, if (AI) machine means invention technology success. Security or cleaning job can be worked from (AI) machine man in some hours every day in order to reduce the long time working hours cleaners or security workers, e.g. one (AI) machine man works 4 hours for cleaning or security job, one day as well as another cleaner or security labor only needs to work 8 hours one day. So total security or cleaning employers can employ 12 hours machine cleaners or security workers and human cleaners or security workers in one day. For long term benefit, Hong Kong or Japan every security or cleaning worker does not need to work 12 hours minimum working hours one day. They won't feel tried and bore and without private with whose families, so who will accept to do these cleaning or security jobs, even they can raise work efficient and performance when who feel happy and health.

So, (AI) technology of machine man invention can raise low skillful labor efficiency and it can help them to avoid abnormal working hours demand in some busy work life countries, such as Hong Kong and Japan. Before, one Japan female labor feel unhappy to work, due to who often needs to work abnormal working hours for her employer and who has less sleeping and without any private time to enjoy her life with her families every day. So this abnormal working hours factor causes her to do commit suicide behavior, then she is die unlucky. So (AI) technology of machine man invention ought avoid abnormal working hours demand for employer in any countries in the future.

The most important occurrence to any employers, some researchers had attempted to do one experiment to find that private research and development , venture capital and public research and development investment all have strong net effect or economic growth with venture capital funding further having the strongest such effect from (AI) technology. The researchers hypothesize the venture capital investment contributes to economic growth through (AI) technology innovation and by the capacity of an economy to use existing (AI) technology knowledge to increase productivity. They predict the impacts of venture capital, business-research and development and public research and development can raise multi factor productivity from (AI) technology introduction.

Can (AI) technology influence the economic development to developing countries? The developing regions of the world contain most of natural resources. If one day, (AI) technology has invent one kind of machine man which can assist any gas or oil workers to seek any new oil/gas natural resource locations easily. I believe that (AI) technology can help these natural resource exploitation countries will gain economic benefit more easily. So, (AI) driven technology can be used to change to create any new opportunities to address poor management or resources and improve human well being, such as Africa Latin America and India can use (AI) technology machine man to seek any oil/gas natural resource countries exploitation activities to attempt to gain much economic benefits.

- Why will (AI) technology grow economic

development ?

Nowadays, increases in capital and labor are no longer driving the levels of economic growth, such as (AI) technology. The ability of increase in capital investment and in labor of traditional drivers of production, have no longer to be enjoyed in most developed economies ,e.g. developed country, US, UK . However, artificial intelligence has the potential to overcome the physical limitation of capital and labor to avoid missing out on this opportunity. So, policy makers and business leaders must prepare for and work toward a future with artificial intelligence. They must do with the idea that (AI) is another simply method to enhance productivity method . Rather they must see (AI) as the tool that can transform thinking about how growth is created.

Economists have always thought of new technologies are as driving growth their ability to enhancing. It can replace labor and capital factor of production. So, it brings this question: What is the factor of production (AI) technology characteristics. They key factor is to see (AI) technology as a capital-labor .

(AI) can replicate labor activities at much greater scale and speed, and to even perform some tasks began the capabilities of human. For example, by using virtual assistants , 1000 legal documents can be reviewed in a matter of days instead of taking three people six moths to complete. Some (AI) technology may be one kind of factor of production in the future. For another example, people will work in workplace digitalization environment. So, in the future, working environment and information management are automated. Such as Konica camera sale company will use workplace digitalization. So , (AI) technology can provide workplace digitalization in order to raise productivity efficiency. (AI) technology will be one kind of production which is replaced by workplace digitalization and it will grow any organization productivity efficiently. Then, (AI) technology will assist overall social economy growth , due to productivity is raised and products can be produced in short time to prepare to sell in consumption market. So, time will be shortened to increase GDP growth fast for the development of (AI) technology countries.

- How can (AI) technology impact to global economic and social and psychological changes?

What will be the development of (AI) technology and predictions concerning the future evolution? The computers and robots will develop conscious, intelligent and minds into humans, enhancing psychological and behavioral abilities and allowing for direct communication with (AI) minds. (AI) technology will be impacted human life by (AI) technology information communicative and environmental influence. A " world brain" and " world mind", this psychological system will be enhanced and enriched the capacities of both individual and collective cognition by (AI) technology of service industries.

(AI) technology with influence these human needs of service industries changes, such as , biological science, finance, entertainment, business, biological science, transportation, communication military etc. The personal computer evolution, the internet and the world wide web which exploded on the scene, linking business, homes, schools, social organizations which were a completely unpredicted phenomenon to influence human life. Kurzweil (1999) predicts that by 2029 year, most human communication will be with machines. According to Person, by 2100 year, there will be human machine convergence.

How can (AI) technology influence environmental protection to make benefits to farming economic growth? (AI) technology can be applied to predict how to solve environmental pollution challenge to avoid to damage any crop or vegetable or rice or fruit etc. food growth. Because environmental experts can gather global environmental pollution data from an environmental database to build a perform a systematic analysis from (AI) technology. The first step is this broad analysis can include understanding, statistical and data gathering techniques to obtain the relevant data, the correlation among the variables involved, and a list of possible models. The next step is to select a set of methods and models that cover all kinds of knowledge and functionalities needed for the decision making process. Once the models are selected, they must be fully implemented by means of machine learning , data mining, statistical or numerical technique. After that, those models must be integrated to build the whole EDSS. The EDSS must be tested to check its performance, accuracy, usefulness and reliability, both from the user's and (AI) technology/computer scientist's point of view. If these is any wrong feature in any development stage, such as model's integration, models' implementation, selection of models, database, problem analysis etc. the developers must come back in the update th required components. When the evaluation phase is all right, the EDSS is ready to be applied to the environment. The great contribution of artificial intelligence to EDSS the integration of several methods complementing the classical statistical models/simulation , statistical analysis, linear models, etc. and numerical models (control algorithms, optimization techniques etc.) .

This cooperation makes the resulting systems more reliable and powerful in coping with real world environment systems. Date interpretation has been a principal area of research in (AI) technology since the very beginning. The most demanding problem in the environmental assessment context. Knowledge representation permits the definition of the different types of data that the existing methods adapt to the process. There is also a lot of work to clean, repair and transform the huge available quantities of raw data. Apart from this, the availability of meta-information or background knowledge is required to guide the process. Data mining is multi-disciplinary: It covers expert systems, data based technology, statistics, data visualization and unsupervised machine learning. These techniques operate at the level of data and background information, where numerous and often incompatible new commensurate pieces of information from disparate sources have to be brought together (K, Fedra, 1994).

So, it seems that in the future, (AI) technology with the increasing maturity in particular those related to knowledge and engineering, new dimensions can be assisted to users in environmental decision making are available. For example, many environmental systems are characterized both by incomplete models and by limited data. Hence, in the future, (AI) technology will be applied to predict climate change to reduce crop or fruit etc. food agriculture challenge by climate change bad influence.

● Will (AI) technology influence digital economy change to manufacturing industry ?

To understand how the manufacturing business must adapt to prosper in the technology, we need to understand how (AI) technology will change us to shape our daily habits to satisfy our expectation of products to how we shop and even the immediate of the entire process. For example, taxi services are in the crosshairs as on demand transportation services like, available of the touch of a smart phone button expand. In fact, Yellow lab, US country , san Francisco city's largest taxi company is filing for bankruptcy as the industry starts to change faster than almost anyone expected. However, at this point, its more than an app that is changing, some our taxi passengers renting taxi transportation to catch consumption behavior.

(AI) technology will influence digital economy for taxi passenger's individual customer experience, offering a growing renting taxi to catch of service and feedback opportunities when any one taxi passenger who chooses to use mobile phone app online tool to prepaid to rent any taxi more easily.

Also in the long term, (AI) technology can influence vehicles drive themselves of behavior. Already, companies like Google and GM are working on projects to bring fleets of autonomous vehicles to cities at the path of a button.

Moreover, this on-demand service model is beginning to appear across a much broader range of markets. For example , Amazon company is investing in its own fleet of trucks, planes and even drone at the same time as it pushes for same-day delivery of products. As some point, vehicles will be autonomous too. So, it seems that (AI) technique will influence any transportations choose to use digital autonomous driving technology in the future . For Amazon company case, it is not stopping of logistics. It is also aiming to automatically manage the supply of consumer home products with its recently launched Amazon replenishment service, Dash. Dash is a digital service that enables that connected derive to automatically order physical products from Amazon when supplies are running low. So, it seems (AI) technology will be applied to logistic function by digital technology method introduction in the future.

Hence autonomous vehicles will optimize industry supply chains and logistics operations through increased efficiency and flexibility. In fact, fully automated and lean supply chains will keep reduce load sizes and inventory by leveraging smart distribution technologies and smaller autonomous vehicles by machine man assistance. If Amazon continues to grow market share for online sales by reducing effort required by the consumer to place an order, when also contributing the almost immediate delivery of products to the doorstep. So, it will further fuel the trend toward on-demand derive. As Amazon company fuels the on-demand economy, consumers will expect immediacy in more parts of the digital economy. On top of speed, consumers increasing expect more personalization options.

So, (AI) technology will influence digital manufacturing, such as Amazon publishing to monitor every aspect of every process in real -time and communicating to self-optimized deep learning robotics, new methods of high volume and high customization will become possible. Then, as products merge into product platforms and even services, manufacturers have the opportunity to provide components and platforms used by smaller players. So, (AI) technology will influence manufacturing industry to choose automated SMI lines, robots installed, automation engineers.

Another future (AI) technology development can be applied to space science aspect, such as Automation engineering space in manufacturing process to achieve digital manufacturing benefits to any businesses in the future. Such as reducing cost, shortening manufacturing time, raising efficiency, shortening delivery products to client individual time. How can artificial intelligence give the need and advanced fast and evaluation methods benefits for space exploration? When US NASA (space exploration organization) achieves any space exploration missions, it will answer this question:

When is it useful to have a machine use (AI) technology to achieve a decision? After all, after millions of years of space exploration and rough 10,000 years of civilization, humans are usually quite good at making decisions in complex uncertain environments. Through, Johns Hoplains University's Applied Physical Lab. Research in (AI) technology enabled systems, which has identified three general use cases for (AI) technology to explore space mission:

First, for some tasks (AI) technology is more cost effectiveness than human. Second, (AI) technology is better suited than humans at solving some, but not all problems. Third, (AI) technology allows NASA organization's space exploration mission to develop machines that ate capable of responding faster than when a human is in the decision loop (D. Scheidt, 2012, A. Castano et. al. 2008).

So, the use of (AI) technology to enable science by observing the pace of rapidly evolving phenomena was demonstrated. It is more effectively coordinating and (AI) technology utilizing to earn economic benefits to use for space exploration mission.

However, (AI) technology also have current risk for space exploration. Today (AI) technology is immature and requires further development to reach its potential. For instance, the (AI) technology algorithms that detected the dust derive could not have identified whether the Martain weather represented a threat to the cover. Also it can not yet use instrument input to determine what, where and how to autonomously make the next space science measurement. An equally important factor limiting (AI)'s deployment is that lacks the methodology and technology to effectively test (AI) technology. So, the challenge will testing (AI) enabled system is how (AI) performance can be measured. It would be NASA organization's difficulty to find (AI) technology to develop to carry on researching any space exploration missions in the future. However, (AI) technology will be a good economic benefit choice for space exploration mission in the future.

- What is artificial intelligence potential
benefits and ethical considerations?

The ability of (AI) technology systems to transform vast amounts of complex information into insight has the potential to help solve manufacturing or service challenges for human needs. However, to reap the societal benefits of (AI) systems, humans will need to trust then and make sure that which follow the same ethical principles, moral values, professional codes and social norms that we humans would follow in the same scenario, research and educational efforts as well as carefully designed regulation in order to achieve the most effort of economic benefits goals. For example, international business machines corporation (IBM) is actively engaged both competitors , in global discussions about how to make (AI) ethical and as beneficial as possible for people as social economic benefits. (AI) is usually defined as the " capability of a computer program to perform tasks or reasoning processes " that human usually associate to intelligence in a human being. Often, it has to do with the ability to make a good decision, even when there is uncertainty, too much information to handle. As an example, play chess or complex card games of entertainment activities is believed to need some form of intelligence in a human being, as well as choosing the best medical facilities in a difficult medical case, or creating something new, such as mathematical theorem or even some form of act, or even driving automatic machine man (self driving vehicle) replacing human driving in the middle of a crowded city.

(AI) needs depends on what we consider being intelligence in the behavior of a human being act a certain point in time. If human belief about human intelligence changes and we don't believe any longer that a certain task requires intelligence, then a computer program performing that task is no longer part of (AI), it becomes just another boring computer program. So, it means that (AI) technology will replace some old computer programs, if human can invent new generation of (AI) software for any functions or activities to satisfy human needs.

As IBM, it argues intelligence. This means that we aim to build systems that enhance and scale human expertise and skills rather than replacing them. We therefore focus on practical applications of (AI) capabilities that assist people in performing well-defined tasks of needs by exploiting and wide range of (AI)-based services. We also use the term " cognitive computing" it is mean a comprehensive net of capabilities based on technology. It comprises the fields of machine learning, reasoning and decision technologies, language, speech and vision recognition and processing technologies, high performance and high efficient functions for any industries or individual consumers needs. For example, robotics, which are usually very good at doing what which are supposed to in any environment, much have public shopping center, factory etc. places which need simply services from the robot (machine man), such as cleans the floor of our houses to the robot that can work together with humans in production chains, passing through the warehouse, robots can take care of the tasks of an entire warehouse and the companion robots like Nao, Pepper, Aibo and Giraff, who can entertain use, talk to use and help elderly people to stay connected to their friends, relatives and doctors.

Google company is building automatic machine (self-driving cars) and has acquired more than 10 robotics companies. Facebook had opened whole new research facility only on (AI) research. Apply computer has developed Siri. Microsoft computer company has built a similar personalized assistant. Google has Deep mind, a UK company whose long term aim is to build general (AI) and has already great potential to win game to the world champion and IBM is investing a huge amount of resources in applying its Watson cognitive computing system to the medical domains to finance and to personalized education. In Europe, IBM is establishing new centers in Munich and Milan focused in the application of cognitive computer capabilities to the internet of things and healthcare respectively.

For example, automatic machine man (self-driving cars) are all about (AI), which used to be able to see what happens in the street (signals ,lanes, other cars, pedestrians, traffic lights, which need to able predict what other cars and pedestrians will do, and who need to be able to cope with unforeseen situations. Since, most car accidents are due to human fault, it is estimated that the adoption of self-driving cars will save about half of the lives that are usually last in car accidents.

IBM Watson company has to understand spoken language, make sense of massive amount to text , respond correctly to questions in many categories, as well as assess its own confidence in responding to such questions. In the future, (AI) technology can own question/answering capabilities that would be very useful, for example, in assisting a doctor when trying to some to the correct diagnosis for a patient and to propose the best therapy .

Intelligent machines can also rely on huge amounts of data to be used to learn how to make better decisions. This data comes from all of us over the years Facebook users have uploaded more than 250 billion pictures and every day who upload about 350 million more. Every second, we submit 40,000 google search queries. So, (AI) technology will be connected through the web from appliances to traffic lights from cars to watches. Other tasks that are very easy for humans are physical and manipulation tasks, such as walking , running, picking up an object to make its shape and location, restricted environment. But (AI) machine man technology still not able to have the general physical and manipulation capabilities even of a 6 year old.

So, it brings this question: Why do (AI) scientists need to concern ethics? Because (AI) technology is complex, information into insight has the potential to reveal long held secrets and help solve some of the world's most difficult problems. (AI) systems can potentially be used to help discover insights to treat disease, predict the whether, and manage the global economy. So, ethic issues is important to and (AI) scientists . If any one new (AI) technology research investigation could success, it will be a secret to and the (AI) scientists can not permit to their loyalty to any competitors to damage the fair (AI) technology products trading market. The country (countries) (AI) technology scientists need to concern ethic issues, who need to keep secrets for their countries economic or/and social benefits.

This is moral issues to any countries/country loyalty is whose countries intangible assets. They can not sell (AI) loyalty to any their countries to assist whose economic benefits immorally.

- How can (AI) technology influence to global
health care economy development?

According to (AI) lecturer analysis, when combined key clinical health (AI) application can potentially create $150 billion in annual savings for the US healthcare economy by 2026 year. (AI) technology is re-winning modern conception of healthcare delivery. It enables machines to sense, comprehend, act and learn. So which can perform administrative and clinical healthcare functions (Accenture, 2017).

It will help health care service organizations to reduce health care cost, will improve and raise service quality and access. So, (AI) health market size will be predicted growth. (AI) applications in health care include robot-assisted surgery, virtual nursing assistant, administrative workflow assistant, fraud detection, error reduction connected machines, clinical trial participant identifier, preliminary diagnosis, automated image diagnosis and cybersecurity.

What kind of benefits (AI) technology can contribute to healthcare service? (AI) technology can deliver what many health care organizations need, such as financial and operational of labor costs, digital expectations from patient consumers how to use (AI) technology to solve interoperability challenges in any healthcare organizations. Also (AI) technology can be applied to wellness an d lifestyle management, diagnostics, delivers financially but also way of organizational and workflow improvement. So, (AI) technology will be continue to become most prevalent and adoption to healthcare organizations , which must need to enhance structure to be position to take full advantages of new (AI) technological capabilities. (AI) technology can change the nature of work and employment is rapidly changing to make the best use of both humans and (AI) talent in healthcare industry in the future. For example, (AI) technology offers a way to fill in gaps and the rising labor shortage in healthcare. According to Accenture analysis, the physicians shortage is increasing. However, (AI) technology will manufacture healthcare machine men to replace physicians in future one day(2017). Hence, (AI) technology will be invented to raise health care service staffs work efficiency and performance in any hospitals or clinics in the future.

In conclusion, (AI) technology will raise efficiency for any service or manufacturing industries in the future, although, it is possible that it will also rise low skillful workers unemployment numbers. But, the most important influence to human technological innovation will be risen and it will influence human life will be changed to be better, e.g. self drive cars, health care physician machine men, machine man cleaners etc. intelligent machine men will be manufactured to serve for our daily life. Furthermore, (AI) technological products will influence countries trading, some low technological development countries manufacturing businessmen can choose to buy any (AI) products to raise whose productivity and efficiency and reducing cost to achieve economic cost saving result. Also, GDP of trading growth income will increase to the (AI) products sale countries. Hence, it will be beneficial to economic development to both developed and developing countries both in the future as well as (AI) scientists time and money spending will be valued to continue to invest (AI) technology development for human life and economy benefits for long term.

In conclusion (AI) technology will raise macro economy growth and it can create many (AI) jobs , but it also raise the low level technological worker unemployment change. In the future, (AI) technology can be applied to digital technology to attempt to invent any new undiscovered (AI) and digital technology. So, it needs any scientists to continue to research how digital and (AI) technology can be mixed to satisfy human's future undiscovered needs.

Reference

A. Castano et. al. " Automatic detection of dust devils and clouds at Mars" Machine vision and applications, Oct. 2008, vol. 19, no 5-6, pp. 467-482.

Accenture, " Why artificial intelligence is the future of growth"(2017) <http://www.accenture.com/us-en/insight-artificial-intelligence-future-growth>.

D. Schedidt , Unmanned Air Vehicle Command And Control, Handbook Of Unmanned Air Vehicles, Springer-Verlag, 2014. Facebook (AI) Research Available at https://research.facebook.com/ai, research at google, machine intelligence available at
http://research.google.com/pubs/machineintellige nce.html; micro soft research-machine learning and artificial intelligence available at http://research.microsoft.com/en-us/research- areas/machine-learning-ai.aspx.

International Federation Of Robotics, 2016. IFR press release world robotics report. IFR, org . 29 Sept. Accessed Feb. 01, 2017. http://www.ifr.org/news/ifr-press-release/world-robitics report -2016-8321.

K, Fedra , "GIS and environmental modelling" in environmental modelling with GIS, edited by M.F. Goodchild.B.O. Parks and L.T. Steyaert, Oxford University press, pp. 35-50, 1994.

Keynes, J.M. (1933). Economic possibilities for our
grandchildren (1930). Essays in persuasion, pp.358-73.

Mckinsey & Company (2013, May). Disruptive
technologies: Advices that will transform life,
business and the global economy , USA.
Ministry of economy, trade and industry, Japan, 2015, Japan's robot strategy. Ministry of economy, trade and industry.

Ray Kurzweil , The age of spiritual machines (1999) is cited numerously through this chapter: Kurzweilai.net http://www.kurzweilai.net

Rich, Elaine & Knight, Kevin, Artificial Intelligence Second Edition, 1991, New York; Mc-Graw-Hill.

THREE

AI INFLUENCES HOUSEHOLDER ENERGY USING BEHAVIOR AT HOME

House quality influences householder electricity
energy consumption behavior

Can apply artificial intelligence measure the house quality how influences the householder electricity energy consumption or useful activities to be more or less? In general, house owners have both intentions for whose property. One intention is living the house by householder himself or herself or householder with families themselves. Another intention is that renting to others to receive rent income (landlord). So, in the housing market, the housing consumer includes either the property owner intents to rent to others to live for rent income aim or the property buyers intents to buy the house to be house owner to live. Does these both different property purchase intentions, which will influence the householder's attitude to use electricity energy consumption desire to be more or less, due to the householder's demand to whose house quality factor influence? This is one interesting question concerns the householder electricity energy consumption desire change to the householder, due to house's investment or house's living intention influences to house quality factor.

How does house quality factor influence to householder electricity energy consumption desire to be more or less? Has it relationship between house quality and house investment or living intention to cause house quality demand to influence the householder electricity energy consumption desire change or demand to be more or less? Has it relationship between regional housing market living or rent investment intentions, housing quality and electricity energy consumption more or less desire? I suppose that the determinants of the residential electricity energy demand form space-heating and cooking, due to the property quality demand influence and the householder's living or rent investment intention influence both, which will influence the householder's electricity energy consumption or useful behavior when he/she/they is/are living in the house.

I argue that rent properties are not only consumer goods, but it also constitute financial market assets. It is therefore reasonable to assume that rational (rent income investment intention) investors choose to raise housing quality (e.g. thermal insulation technological installing at home, heating or cooling technology or artificial intelligent window, lighting, door opening or closing) in order to attract many people choose to rent whose house to live. The householder's aim is to achieve an acceptable return on investment (ROI) or raising rent income aim when he/she rents whose house to anyone, it is easy to attract many people to choose to pay higher rent his/her house to live in the property rent market. Moreover, the another important factor is that rents and future house sale prices of properties differ regionally (or even locally), and largely depend on housing market fundamentals, such as either the house living buyer's income levels or the house rent buyer's income levels, vacancy rates, and/or householder investor's expectations.

Thus, if the householder expects to rent whose house and raises rent to attract many people choose to rent whose house to live, who will attempt to install many new technology in order to satisfy their high quality of life need when they can pay higher rent to rent to choose to rent whose houses to live. Their aim only achieves to raise housing quality, but any new technology will lead to increase electricity energy consumption or use in the house.

Hence, any high quality of houses will influence the householder to use or consume more electricity energy at home. It means that the householder will choose to consume or use more electricity energy at home, if he/she or the family householder demands to live more comfortable house and he/she/they can have high quality of living life at home. This comfortable living demand to the householder (property renter or property buyer) view point can explain why the better quality of house factor will influence the electricity energy consumption desire to the householder also to be more daily.

I shall indicate one home electricity energy consumption experiment, it indicated that utilizing aggregate data on regional space-heating energy consumption form over 300,000 apartment buildings in 97 German planning regions. The study applies structural equation modelling to estimate the influence of housing market fundamentals on the level of housing quality, and subsequently on regional electricity energy consumption. Consequently, it suggests that housing market fundamental explain regional differences in the housing quality.

In particular, findings show that the level of per capita income, investor' expectations about future housing market development as well as vacancy all explain regional differences in housing quality has a significant impact on electricity energy consumption.

In the way, this experiment can indicate evidence that regional housing market fundamental have a substantial influence on regional levels of housing quality and energy consumption desires to the German regional householders. This Germany regional householder experiment found important implications for high or low housing quality of the regional property building and householders either property living or property rent intention of comfortable living feeling need factor which will influence the regional property householder electricity energy consumption desire to be raised or reduced. These factors will influence the consequence of electricity energy demand to be increased or decreased needs every day for the regional householder as well as the country's electricity energy supplier(s) can gather the regional properties whether they are high or low quality to predict the regional properties householders' electricity energy consumption supply budget more accurate. It implies that an important determinant of residential housing quality will have possible to influence electricity energy demand to be more or less for long term. IN particular, this Germany regional residential experiment can explain and find an important role in formulating assumptions about the quality factor has chance to influence the regional residential future levels of electricity energy efficiency and consumption in the country. Hence, housing developments and electricity energy firms can follow this regional residential housing quality factor to evaluate whether the regional housing market is the corresponding investment patterns as well as the energy researchers can follow the regional residential housing quality whether it is high or low housing quality factor to evaluate the more accurate models of regional electricity energy demand to any regional residential householders' houses in the country.

In consumer behavioral view point, it explains that if the country government expected many householders feel to need to spend much electricity energy or have much electricity energy useful demand or desire at home. The country government ought to encourage the country residential property or house developers choose to build many houses which have technological product installed to satisfy the regional householders' residential comfortable living need when they choose the regions to build the high quality houses to let them to live. Then, the regional householders will be influenced to consume or use much electricity energy at homes, due to they feel that they are living at high quality and comfortable and high building technological installed apartments in the country's regions. Then, the country's government and electricity energy provider(s) may be raise much electricity energy efficiency and supply and profit , due to the regional residential householders' electricity energy consumption or useful desire need is therefore influenced to be more by the regional high quality of residential houses factor. So, the regional high quality of residential house factor will have relationship to the regional electricity energy consumption and efficiency to the regional householder's houses.

Otherwise, if the country government felt electricity energy is shortage, it ought encourage the property developers build many low quality and low building technological houses to let householders to live themselves or rent to others to live in the country's different regional residential development market. Due to the low quality of properties and low technological installed to properties factor which will influence any these different regional residential householders to choose method to solve shortage of electricity energy challenge to the country.

In conclusion, to apply consumer behavioral economic theory to property development market, if property quality factor can really influence the householder's electricity energy consumption desire to be used more or less at home daily. The country's property developers can apply this factor to predict property consumer individual property buying consumption behaviors more accurate. For example, if the US property developer planned to build low quality and low technological design buildings and lesser comfortable residential houses in the region in US. Then, its residential householder target will be trended the less acceptable of electricity energy consumption property buyers to choose to buy these regional properties to live in the US region because they can only accept to spend less electricity energy to use when they are living in the houses in order to save money daily. SO, the low quality , less comfortable and low technological installed design residential houses will satisfy their living needs. Otherwise, if the US property developer planned to build high quality and high technological installed design buildings and more comfortable residential houses in the region in US. Then, its residential householder target will be trended to the more acceptable of electricity energy consumption property buyers to choose to buy these regional properties to live in the US region because they can accept to spend more electricity energy to use when they are living in the houses in order to improve their living of quality. So, they wont's consider to spend more expenditure to use electricity energy for any technological products are installed in their properties in order to satisfy their comfortable living needs at their homes every day.

Consequently, property developers can attempt to gather marketing research concerns whether how many people who accept to use more electricity energy or use less electricity energy in order to predict they ought build how many high quality or low quality houses number in different regions more accurate in themselves countries or overseas countries property development market.

Environmental impacts of householder greenhouse gas electricity energy consumption activities

Can artificial intelligence measure whether what environment factors influence householder electricity energy consumption activities? Has environment factor relationship to influence householder electricity energy consumption behaviors? Socially, householder electricity energy consumption provides us with sources of living satisfaction , but if any sudden environment factor changes, whether it will influence householder consume or use more or less electricity energy decision at home. However, I assume householder electricity energy consumption will have a considerable proportion of the environmental impacts be influenced by our way of life and our economic decision of electricity energy consumption behavior.

What different environmental factors will influence householder electricity energy consumption decision? The external environmental factors include, for example, the country's electricity firms or government changes to electricity energy regulations, electricity energy production technologies change and business practices and government policies changing etc. different external environmental factors will influence any country's electricity energy consumption to householders' consumption desire to be more or less. It will also require changes to influence the householders to consume which kinds of electric products which are needed to be used in different electricity energy natural manufacturing resources.

Why does these external environmental factors impact householders' any behaviors to influence them to concern to use more or less electricity energy power or which kinds of electricity energy products choice at homes. How any why environmental factors impact will influence householder activities at home, such as electricity energy consumption and choice? What are the key components of external environmental factors influence householders' electricity energy consumption behaviors. I shall explain as below:

Firstly, we need to know whether what external environments are which can influence why and how householders need to change their activities to choose more or less or which kinds of energy power to be provided to them to use at home. Who is householder? Householder is an individual, family, or group of individuals living together as unit in a home. Consumption of electricity energy at home may be cooking food needs, needing have colder feeling to turn on fan or air condition at home in summer or needing have warm feeling to turn on heater at home in winter, watching television programs or listening music , playing computer games or used computers activities , reading activities and applying artificial intelligent technological tools to help householders to open or close homes' windows, doors etc. different home equipment which need to use electricity energy provisions. SO, their home activities need to turn on lighting electric tools , televisions, music machines, radios etc. different equipment which need to use electricity energy provision at home. SO, the purpose of householder consumption means consumption by individuals living in a household and it includes consumption both in and outside the home. Why does environmental impacts link to householders' electricity energy consumption? I shall focus on discussing of greenhouse gases (GHGS) energy product how any why it can influenced to householders to use.

The environmental impacts will influence this kind of greenhouse gases (GHGS) energy in the product lifecycle or delivery of the service to link the householder's energy consumption at home such as these several aspects:

Extraction and greenhouse gases production (supply number), physical distribution (delivery far long or close near short distance between the greenhouse gases manufacturing factory and the greenhouse gases supplier), resources consumed by marketing and retail activities (householder's needs to use the quantity of the greenhouse gases energy product), the greenhouse gases consumers search and purchasing activities(e.g. travel to shops, internet purchasing channel, , finding the which kinds of greenhouse gases products from internet, magazines, newspapers, radio advertisements etc. different medias,) , post-use greenhouse gases energy disposal (resale, reused or rubbish). The householder's physical behavioral impact environmental factor will influence how and why he/she chooses to consume greenhouse gases energy daily , e.g. impacts of a housing development, or a wind –farm that supplies greenhouse gases with power. So, the householder's greenhouse gases energy consumption behavior which will depend upon individual personal and subjective perspectives and value.

So, the householder's useful behavior or attitude of greenhouse gases energy product which will influence how he/she/ the family use or consume greenhouse gases energy, such as the householder individual environmental protection attitude which can impact how he/she/the family spends the quantity of greenhouse gases energy every day at home, if the householder does not expect our air or water or land is polluted , due to extraction of any natural gas resources to be manufactured any kinds of greenhouse gases products. Then, this environmental pollution issue will influence some householders choose to reduce to use more quantity of greenhouse gases products every day. Another environmental factors include the bio relates the (unsustainable) use of resources to avoid wasting much greenhouse gases energy to cause greenhouse gases energy supply shortage, avoiding the cause negative impacts of quality life , e.g. noise causing when the extraction of any natural resource from lands to the householder's house is near to the natural resource extraction land and health impacts, e.g. when the greenhouse gas householder user who often use the kind of greenhouse gas product when it is used to cook or heat any equipment to cause they to breathe dirty air at home often. These impacts can be measured in different ways include: monetary costs or loss, physical quantities of resources used or waste or pollution produced and the burden the greenhouse gases energy place on environmental resources. All of these external environment factors will impact the householder individual attitude or behavior how to use or consume greenhouse gases energy product at home.

All these environmental factors concern householder greenhouse gases energy consumer individual consumption attitude is influenced by environment pollution, greenhouse resource supply shortage challenge, greenhouse gases influence the householder's negative quality of life, negative health impacts, noise, waste money , raising economic cost to the householder which will impact whether how the householder choose to use the quantity of greenhouse gases product or the kinds of greenhouse gases products or other kinds of electricity energy products.

However, these are other external environmental factors which can impact how the householder decides to use greenhouse gas product at home. They include: the changes of energy regulation, e.g. the country government has quota number implementation to prohibit to import above the limited quantities of any kinds of greenhouse gas products to any countries. So, when the greenhouse gas energy supplying quantity is decreased, but if the country has may householders who need to buy different kinds of greenhouse gases products to be used at home. Then, the different kinds of import greenhouse gases energy products prices will be raised in possible, due to demand is more than supply in the country's greenhouse gas energy product market. Consequently, if the greenhouse gas energy price us risen above the general social acceptable level to the home greenhouse gas energy product householder consumers. Finally, it will influence them to choose to buy other kinds of gas energy products to replace the greenhouse gas energy product to use at home.

Another side, if the country's greenhouse gas energy manufacturing supplier sudden changes its greenhouse gas energy production technologies to choose to concentrate on manufacturing other kinds of energy products. Then, the greenhouse gas energy supply quantities will be only decreased, even future one day , it will cause greenhouse gas supply shortage challenge to let the country's home greenhouse gas householder consumers who can not buy enough quantity of any kinds of greenhouse gas energy products to satisfy their electricity needs at home every day. Consequently, when future on day , the country greenhouse gas energy manufacturer has none any quantity of greenhouse energy products to supply to the country's greenhouse energy householders to use at home. The, they must only choose other kinds of new energy products to replace the traditional useful greenhouse gas energy products to be used at homes.

In conclusions, these non-controlled external environmental factors can impact and influence the country's every householder consumer individual attitude or consumption behavioral change to how any why the country's householders either choose to buy much or less quantity of greenhouse gas products to use at home.

The effect of house space occupancy
and building characteristics on
householder electricity energy use

Can artificial intelligence mesaure how much space occupancy is the suitable size as well as what the most suitable building characteristics influence each householder electricity energy useful activities or behavious? In general, society believes large space size occupancy house building characteristics factor which will influence householder use more energy at home, e.g. in summer, when the householder is living at the large space size occupancy house, who ought turn on all air conditions or fans at sleeping rooms or eating room or studying room. So, if the householder's house has two to three or more sleeping rooms. Then, he / she needs to buy more air conditions or fans in order to let all rooms' temperature to be fallen down to let he /she feel more cool comfortable feeling when the temperature is above 30 degree or more extreme hot in summer weather. Otherwise, when the temperature is low, e.g. between 0 degree to 10 degree or below 0 degree in winter weather. When the householder is living in one large space size occupancy appartment, which has thee to five sleeping rooms , even more and two studying rooms and one eating room, even more as well as every room has one heater. Then, he / she must turn on all heaters to let who to feel warm feeling when he / she is staying in the house. It brings these interesting questions.

Will large or small size space occupancy housing characteristics influence any householder often turn on heater or air condition or fan in whole house space occupancy area in order to the householder feels warmer or cooler feeling when he /she is staying in the house?

Has any space occupancy housing characteristics relationship to influence any householder to turn on heater or air condition or fan in whole house space occupancy area in order to the householder feels warmer or cooler feeling when he /she is staying in the house?

Does it bring positive relationship between turning on long time fan or air condition or heater and the house occupancy space characteristics is large or small size?

I shall attempt to give psychological evidences to explain the householder's house space occupancy area large or small size factor whether it can influence the householder choose to do long time or short time turning on heater or air condition or fan behavior in order to let he/she/the family to feel more cooler or warmer comfortable feeling when he/she/the family is staying in the house in summer or winter weather.

Does the house occupancy space size characteristics factor is the only one or important factor to influence the householder choose to turn on long or short time fan or air condition or heater in the house to let him/her/the family to feel more cooler or warmer comfortable feeling in summer or winter weather?

I feel that it is not exact right , due to the householder's house space occupancy size whether it is large or small characteristics to influence the householder choose to turn on long time or short time fan or air condition or heater time to let him /her/ the family to feel more cooler or warmer when he / she / the family is staying at home in summer or winter weather. The reason is because that the lifestyle of living quality need is different between developed countries and developing countries. The lifestyle of living quality factor will change the country's householder's expectation about the quality of living life. For example, for Africa, Korea, China , Japan, Hong Kong etc. developing countries. On the lifestyle of living quality need to these developing countries' householders aspect, that will cause a high environmental burden when they need to often turn on air conditions to satisfy more cooler feeling when they are staying at homes in summer or they need often to turn on heaters to satisfy more warmer feeling when they are staying at home in winter. Due to if their houses are large size space occupancy characteristics and they have more than at least two sleeping rooms and studying rooms and eating rooms and toilets number. Then, these householders who are developing countries' large space occupancy size characteristics houses, they won't like often turn on heaters long time to keep more warmer in their indoor whole space area in winter or they won't like often turn on air conditions or fans long time to keep more cooler in the their indoor whole space area house environment in summer .

The reason is possible because that the developing countries' householder chooses often to turn on their heaters or air conditions or fans long time in their houses when they are staying long time in their houses and their houses space occupancy sizes are very large, it will bring the electricity energy to be used more to these developing countries' householders' large space occupancy size characteristic houses. It means that the electricity fee will be also increased due to they often turn on heaters or air conditions or fans long time to keep their indoor temperature to be more cooler in summer or more warmer in winter. So, it seems that the developing countries' householders are living in the house whose space occupancy have very large size characteristics and more than two rooms house characteristics in the developing countries as above. Then, they won't often choose to turn on heaters or air conditions or fans long time to keep more cooler or warmer feeling in their house whole indoor space occupancy environment when they are often staying at home long time.

Their lifestyle of living comfortable feeling are lesser than the developed countries householders. Consequently, their lesser cooling or warming comfortable demand of living lifestyle factor will change their attitudes to use air conditions or fans or heaters turning on time in order to limit heaters or air conditions or fans turning on time to be shorter than the developed countries houeholders' heaters or air conditions or fans turning on time at homes. Due to the long time turnong on air conditions, fans , heaters at the developing countries' householders' homes, it will cause to spend much electricity energy to lead electricity fee charges to be raised to the developing countries' householders ' homes when they are often staying at homes in summer or winter weather. Hence, the house space occupancy large size characteristics ought not influence the developing countries householders choose to turn on air conditions , fans or heaters long time in order to let them to feel more cooler or warmer at homes in summer or winter weather.

So, the developing countries' house space occupancy large size characteristics householders won't be more acceptable to pay higher electricity energy fee when they are staying at homes at summer or winter weather. Due to they do not often choose to turn on heaters, air conditions or fans long time during they are staying at homes. Otherwise, the developed countries, e.g. UK, UK , France, Germany, Swiss, Singapore, Italy etc. countries. In general, these developed countries' householders' living lifestyle quality needs are higher than the developing countries. So, when the summer or winter weather is coming, if the temperature is extreme cold, e.g. below than 0 degree or it is extreme hot, e.g. higher than 30 degee.

Then these developed countries' householders will easy accept to turn on air conditions or fans or heaters long time at home in order to keep their appartment in door temperature to be more cooler in extreme hot in door environment or more warmer in extreme cold in door environment when these developed countries' householders are often staying at homes long time at night after their day time working time or schooling time. Because these developed counties' householders' quality of living lifestyle needs or demands are higher than the developing countries' householders. So, they won't consider that they will pay more electricity fee , due to they often turn on air conditions, fans or heaters long time to let them to feel more comfortable in cooler or warmer indoor large size space occupancy environment. So, it seems that the electricity energy efficiency will be raised to the developed countries' householders who are living in the house space occupancy large size characteristics and they will be possible to pay more electricity fees during they are often staying at home in extreme hot summer or extreme cold winter weather.

In conclusion, due to the living lifestyle quality need (demand) is different between the developed countries' householders and the developing countries' householders. It will influence the householders' long time or short time spending time on air conditions or fans or heaters indoor space occupancy size characteristics environment in order to achieve more cooler or more warmer feeling in their houses. Consequently, the long or short time of turning on air conditions, fans, heaters for the developed or developing countries householders' activities factor will be more influential to compare the house space occupancy large or small size characteristics factor to influence their cooler or warmer feeling in their houses. SO, the house indoor environment electricity energy consumption efficiency degree to the developing or developed countries' every householder house in summer or winter to the developing or developed householders in summer or winter weather , which is more influenced by the living lifestyle qualty factor to the either developed countries or developing countries householders. Hence, any developed or developing countries' electricity suppliers need to consider the building areas of property development market buyers their living style quality demands (needs) whether their living style quality demands are higher or lesser than the other building areas of property development market, they ought not consider whether the building locations of the houses' space occupation sizes whether they are large or small sizes in order to evaluate the householders will live at the building areas of property development locations ,whose electricty energy spending efficiency more accurate.

How to help low income household
earners to reduce not essential
electricity energy expenditure
spending at homes

Can artificial intelligence measure whether which income level influences low income household accept to use the much electricity energy at home? Has it relationship between the householder income and the electricity energy needs? How to evaluate the subsidies and social tariffs to assist lower income earners to analyze household energy consumption more accurate?

Electricity energy is essential needs for every householder at home, e.g. lighting, cooking power, healthcare, sanitation, cooler or warmer temperature indoor control at home. However, for lower income household earners, it its burden when they need often to use electricity energy to supply power to any home electricity tools to do any acticities at homes. If any these countries' lower income householder earner target can not get the reasonable subsidies to assist them to solve any electricity energy tools' electricity energy poer needs. Due to their lower income leve, it is possible that to knfluence them have enough electricity supply to help them to use to cook rice and food and vegatabe to eat, boil water to drink, turning on light tools to help them to read, watch TV, listen radio, music any entertainment or essential needs at homes at night or morning afternoon time. These lower income household earners will be easy to sick , due to they have no enough electricity supply to help them to use use electric bottles to boil water or cook food to eat. Then they only drink not boiled water or not cooked food to eat at homes in possible, due to they have no enough income to pay electricity fees every month.

Hence, how to evaluate the lower income household earners' electricity fee need (demand) level in order to provide the reasonable subsidies amount to assist every country's low income household earner to help them to pay the reasonable electricity fee which is one important issue to every country's government today. It brings this question: How to evaluate or analyze or predict every lower income household individual or family earner's every month electricity energy demand (need) more accurate?

It is one essential issue to be value to consider to every country's government. Moreover, to the estent that energy subsidies must be essential to be provided by public sources to all low income household earners or that a social tariff may be designed for improving access to energy for certain low income social earner groups. Hence, how to structure the energy subsidies between energy and income levels to be better target, such public mechanisms, and to avoid regressive subsidies unfairly. For example, India and China these both countries ' income poverty and energy poverty population are the large number. So , these both countries' governments need to focuse on more aggregated effects and analyze the effects of rural electrification at the local level on the decrease in energy poverty in rural low income poverty and energy poverty householders. Therefore, every country government needs to point regressivity of the subsidy for electricity. There is room to analyze to what extent low income household earners along the income distribution demand some forms of energy, and to suggest better and fair low income targeting household earners energy subsidies supply policies.

Each government does not only consider energy issues from a social point of view, it also needs have a manner to consider a possible link between energy, hunger reduction, and food security for each country's low income household earners group. So, every government has responsibility to calculate the determinants of different sources of energy consumption at the low income houehold earner level for urban and rural both populations in order to evaluate the electricity subsidies and to test whether every low income householder earner characteristics plays a role in determining energy consumption.

In general, in the use of energy measured as that for cooking, such as LPG reduces the exposure of households to hazardous, increases the consumption of different types of foods and medicines, improves the distribution of time between household memners, enables studys with more light, reduces the use of digital computer entertainment tools at home, and moderates the use of wood as fuel, preventing deforestation. These methods are the best suggestions to help low income householder earner groups to reduce time to use electricity at homes. When they spend less time to use electricity to do any not essential activities, e.g. watching television, playing electric games from home computers, listening music. They only use electricity to turn on light read, to turn on rice cooker to cook, when they feel hungey to eat. Then, I believe that these social low income household earner groups will reduce to pay much not essential electricity energy expenditure at homes. Hence, every country government ought need to persuade low income household earners to avoid to use electricity to do any not essential activities in order to raise electricity energy consumption in long term time.

It will bring less amount of energy subsidies expenditure benefits to every country's government. Hence, the success to persuade any countries' low income household earners to reduce to spend much time to do any electric entertainment activities of consumption behaviors at homes often. This is the most efficient and the most successful energy subsidiary method to help them to reduce electricity energy expenditure when they are staying at homes. Hence, if any country government expected the low income household earners can continue really reduce electricity energy expenditure, they need to learn to do the meaning essential activities which are needed to use electricity at home habitally. Then, they can change their electricity useful entertainment living habit, e.g. using computers to play games, listening music, watching television entertainment habits at homes to cause essential daily needs of electricity useful living habit, e.f. using cookers to cook rice or cook food to eat, turning on lights to read , turning on heaters to bath, turning on air conditions to keep cool temperature or turning on heaters to keep warm temperature at homes. Consequently, they won't need to pay much electricity expenditure at home, due to their waste useful electricity entetainment living habits have changed to do any essential useful electricity activities at homes.

Another kind of method to reduce the determinants of energy demand to the low income householder earners. The governments can persuade them to consider the variation factor can influence their electricity energy expenditure are increased or decreased at homes. It is not the electricity or gas price is increased from the electricity suppliers. It is that their bad living habits of waste electricity or gas to do any not essential activities at homes. e.g. the householder often turn on light tools to read or listen music or watch television in whole night, he/she ought need to sleep at night, but he/she does not go to bed to sleep in whole night. He/she chooses to turn on light to do these activities. Then, he/ she will waste much electricity at whole night. Also, some householders like to bath more than half hour, even one hour, when it is winter, they need to turn on heaters to provide electricity to cause the bath room has warm water to provide to them to bath, Their long time bathing behaviors will be also waste electricity or gas energy from long time heating in bath rooms. So, they need to change their waste electricity consumption living behaviors at homes.

So, I suggest that some low income household earners will need to be taught to change their bad using electricity enery living habits from governments' public relation promotion in order to change the low income household earners' bad or incorrected useful electricity or gas living attitude to achieve and to avoid them often to do electricity or gas energy waste behaviors at homes. So, different countries' governments need to teach them how to do the

correct or right electricty or gas useful activities (living habits) or let them know or feel how to use their electricity or gas which can help them to reduce to waste the not essential extra electricity or gas energy. Consequently, they must reduce electricity or gas expenditure as well as electricity or gas shortage challenge won't be caused by their electricity or gas useful waste behaviors (activities) at homes.
In conclusion, energy subsidies method is not the best solution to help low income household earners to reduce to use electricity or gas energy. Because it is only short term benefit to reduce their electricity or gas expenditure at homes. The best solution is that to let them to know or feel why and how they have responsibilities to change their incorrent or wrong electricity or gas consumption bad habits in order to avoid global electricity or gas energy is waste to be used, even it is caused shortage from householders' energy waste behaviors.

Factors influence householder energy
efficient consumption behaviors
at homes

Can artificial intelligence measure how householder uses energy level at home ? What factors can influence householders how to use energy in efficient way at homes. It depends on different countries householders' living habits to cause their choices to use energy efficiently at homes. In general, global householders energy every day consumption or use aims include cooking, heating, and cooling or warming rooms, lighting , water-boiled use and computer playing games entertainment etc. activities at homes every day. Some activities are often essential at homes, e.g. cooking, cooling or warming temperature in rooms, lighting , water-boiled use. So, their activities must not avoid to use energy at homes often. Otherwise, some activities are not essential at homes, e.g. playing entertainment games from computers, cooling rooms in summer, listening music, watching television etc. these activities. The householder can choose either to use energy to turn on these equipment tools or not to do these non essential activities at homes often. In general, householders rely on energy to make ourselves lives comfortable, productive and enjoyable. However, global householders need to learn how we can use energy resources wisely because global every householder has responsibility to manage resources includes: reducing total energy use and using energy more efficiently in order to avoid energy shortage crise occurrence. The choices are make about how we use energy, e.g. turning machines off when not in use of choosing to buy energy efficieny appliances will have increasing impacts on the quality of our environment and lives.
Energy conservation includes any behavior that results in the use of less energy. Energy efficiency involves the use of technology that requires less energy to perform the same function. For example, a compact fluorescent light buld that uses less energy to produce the same amount of light as an incandescent light buib is an example of energy efficiency. So, a householder's decision to place an incanadescent light bulb with compact fluorescent is an example of energy conservation. So, as individuals, every countries' householder choices and actions can result in a significant reduction in the amount of energy used in each sector of the economy.
So, I bring this interesting question: What factors can influence householder to choose to do any efficient energy

consumption or useful behaviors at homes? I believe every countries' householders will have their different living attitudes and their living attitudes can influence their behaviors or activities to choose hoe to use energy at home. I shall indicate some countries' householders' living attitudes to explain the factors can influence them to use energy efficiency at homes as below:

- Is the low income and rising price of modern fuels both factors best to influence Nigeria householders choose to use energy efficiently?

Firstly, for Nigeria householders energy consumption habit at homes example, it is richly with natural resources, modern energy resources which provide many householders with biomass (mostly firewood) and some other householders modern energy sources, such as kevosene, liquefied, petroleum, gas and electricity for their use. So, it is one country which can manufacture to provide energy for itself to use. It doesn't need to depend on other countries to import any kinds of energy to householders to buy to use at homes. But, it has social challenge, the poverty problem in Nigeria goes beyond low income, savings and growth rate, due to its low level of education, poor governamce, high level of unemployment factors influence.

It is important to know how Nigeria householders meet their basic energy needs between poverty and energy can bde described in terms of quality and quantity of energy used. Generally, most poor householders use biomass fuels because of affordability and they (householders) do not have energy equipment (such as, gas cookers, electric cookers etc.) . So, it seems Nigeria householders won't demand their living quality to be improved. It implies that they will use any kinds of energy efficiently at homes, e.g. gas, electricity, due to they find themselves in energy poverty. Although, this country has enough nature resources to manufacture energy to provide to householders to use, but due to many people are low income group, so they won't spend too much expenditure to buy much energy to use at homes. So, the rising prices of modern fuels, such as liquefied, petroleum , gas (LPG) and electricity and their erratic supply have made many householders revert to the use of traditional fuel, such as firewood and charcoal.

It brings this questions: Is the low income and rising price of modern fuels both factors best to influence Nigeria householders choose to use energy efficiently?

The hypothes is predicated on the economic theory of consumer behavior. However, when income increases, householders not only consume more of the same goods, they also need higher quality . So, it applies economic theory to householder's energy consumption behavior at home. It explains why low living standards induce greater dependence on firewood and other biomass fuels owing to a combination of income and substitution effects, such as Nigeria low income household energy home users case. it explains why Nigeria householders can accept to use firewood and charaval traditional energy to replace liquefied, petroleum , gas (LPG) and electricity modern energy . So, economic theory explains the Nigeria household energy users why they can accept to use traditional energy to replace modern energy and their energy useful or consumption behaviors are efficient at homes. Although, Nigeria has enough natural resource to manufacture modern energy to supply to householders to use at homes. But, due to these modern energy products prices are raised to the price level of householders who can not accept. it causes to Nigeria householders only choose to buy the cheap biomass, firewoods to replace high price of modern energy products to use at home often. So, they can accept their quality of living to be fallen down. So, expensive modern energy product price is one factor to influence some countries' householders to choose to buy cheap traditional poor quality of nature energy, e.g. firewood or biomass, to use at homes. Hence, they can raise energy efficiency to use when they choose to use traditional nature energy to replace modern nature energy at homes.

● Does season factor influence New Zealand householders' energy consumption behaviors at homes

Secondly, for New Zealand householders energy consumption habits at homes , for example, their living quality needs are general comfortable need feeling. Their countries' houses of space heating was found to average 34% of total housholder energy use. The relation to space heating includes low indirect temperature are associated with persistent under-heating , whether some space heating sources tend to be higher or lower in winter indoor temperature than others and winter indoor temperatures are compared to international benchmarks and established healthy temperature ranges. So, New Zealand occupant's perceptions of winter indoor temperature conditions are presented and explored in relation to heating patterns and household energy consumption. So, it seems that NZ winter temperature is low. Moreover, it will influence householders need to turn on heaters to keep more warmer feeling indoor. Then, they will use more electricity energy. In special, if the householders' houses spaces are large sizes . Hence, their heaters need long time to keep whole houses' areas or spaces or rooms temperature to be rised up in order to let they do not feel very cold in winter. So, NZ's winter extreme cold weather will influence householders' energy use or consumption to be increased in winter.

The electricity efficiency to every NZ householder is very high in winter to compare spring, summer, autumn seasons. Hence, if NZ electricity suppliers expected to forecast electricity consumption more accurate in NZ. In order to ease the life for both electric net designers and electricity suppliers, it was decided to find out, how the NZ weather conditions and every householder's house space size factors to influence the power consumption to NZ householders. If there is a clear trend observed , then this relation can be used for power consumption forecasts to NZ householders. Why does NZ weather condition factor and householder's house space size factor can predict householders' electricity consumption at homes. Due to geographic location on the global the lowest south sets specific conditions for weather, such as NZ's south island geographic location is near to south ocean in our earth. It is a country where average annual temperatures are well between 10 degree to below 10 degree at NZ south island special geographic location to near to the sourth ocean in our earth at the same time.

However, large part of mankind is living in the conditions where there are four different seasons in NZ geographic location, dark winter, which is cold and snowy, spring with rising temperature and high precipitation, sunny , dry and rather hot summer, and windy and wet autumn. These conditions lead to different patterns in electric appliances use in NZ householders, in special, in NZ south island householders. If trends in electric energy use have substantial correlation with weather conditions, this can help NZ electric energy suppliers and producers to forecast electricity consumption and thus organize and manage production of electric energy.

Consequently, it will lead to much more stability in energy supply to NZ every householder. For example, when the

NZ energy supplier gathers data concerns every householder's house space size data, e.g. the house has how many sleeping rooms, toilets, bath rooms, eating rooms and reading rooms number, even the house has how many family members are living in every NZ geographical location. Then if it can follow different location of NZ houses spaces sizes whether they are large or small space size as well as whethe every house has how many family members are living to evaluate whether how much electricity efficiency can satisfy their comfortable living needs in winter. Then, it can evaluate whether they will use how much electricity efficiency for their needs in different seasons. If in winter, many householders are living in the large space size house in the geographic location. Then, it is possible that the geographic location is householders will use much electricity efficiency and where geographic location hosueholders who will be possible to pay the most highest electricity fee to compare the other geographic location of small space size of house householders. Hence, weather factor is the most influential to change NZ householders ‘ electricity energy consumption behaviors at homes.

● Urbanization level and income per capita both tangible factors as well as temperature (weather variation factor) will have close relationship to influence China householder energy consumption or useful needs at home every day

For China householder energy consumption habit example, what factors can determine to impact this country's householders energy useful behavior at homes? Can the impacts of these factors be quntified? What are China householder energy consumption trends and characteristics? I shall explan as below:

I believe the influential factors include these three aspects to China householder energy users: Income per capita, urbanization level an annual average temperature (weather). These factors will influence any China householder energy useful or consumption behavior at homes.

Temperature (weather variation factor) is intangible from eastern region to western region of Chin, variances largely depend upon economic level and the provincial level. So, some regions were warmer and cooler temperature will influence the regional China householder how to use electricity. In addition, th influence of urbanization level varies according to income level as well as the urbanization level has more significant impact on the structure and efficiency of China householder energy consumption thatn on its quantity. So, the urbanization level and income per capita both tangible factors will have close relationship to influence China householder energy consumption or useful needs at home every day. Moreover, these two tangible factors (urbanization level and income per capita both factors) have the more influential to impact China any one of household family energy consumption or useful habit to compare temperature factor at home. Because temperature can only influence than to choose to turn on heaters to keep more cooler in summer or turn on air conditions (fans) to keep more warmer in winter.

The electricity energy needs for these equopment tools which will be influenced less. Otherwise, the urbanization level and income per family householder how to choose to spend more or less electricity or gas etc. energy at homes. Because in behavioral economy view point, when individual householder has more income and the urban in the China geographic location is lising many high income and high household families memebrs to every house. Then, the urbanization household energy household enery useful or consumption level will be raised. Such as China household electricity users case, e.g. large cities have many high income and many houses have more than four families members to live on one house together. Then, the electricity or gas energy efficiency will be influenced to rise. The city urbanization and per capita income level is high to these large cities have high to income population, who are living in these cities in China.

Moreover, the impact of lifestyle on energy use mainly reflects types and purposes of fuels are chosen by different China households factor which will influence the urbanization level of energy choice use. China is a country with typical binary economics and social diversity and these is significant difference in the consumption pattern between urban and rural regions. Urban residents consume high-quality energy, such as electricity, natural gas , heating power, solar energy and gasoline. For rural residents, usually use coal, and bismass energy because they are cheaper price energy products which requires much time and labor and are heavy indoor pollutants . The difference in energy consumption pattern between urban and rural China residents is closely related related to living of quality needs, building structure, e.g. steel or stone etc. different materials, manufacture, easily access clean and effective feels through the electric grid, natural gas network and district heating systems.

Therefore, it explains why urbanization level is as an integrated variable reflecting social progress situation to influence urban and rural regions, such as large cities , small cities and rural countryside regions' household energy consumption or useful behaviors which have differnet kinds of fuel useful demands and energy efficiencies qualify and quantity demand, or needs at homes. Consequently, it explains, urbanization level and income per captia level both factors are more influential to China household energy consumption at home to compare temperature (weather , seasonal) factor.

● Employment rates or gross domestic product macro economic variation factor, residential space size factor, and the government's implementation of energy labeling schemes provide significant impacts on Taiwan residential electricity consumption .

For Taiwan householder electricity consumption characteristics in the residential sector, which has different factors and pattern to compare China householder electricity householder electricity consumption habit at home. Although, they are the same Asia country. I shall explain these reasons as below:

For Taiwan electricity householder factors influence their energy useful or consumption behaviors at homes. The main factors can influence their electricity energy useful patterns include: employment rates or gross domestic product macro economic variation factor, residential space size factor, and the government's implementation of energy labeling schemes provide significant impacts on Taiwan residential electricity consumption . However, the impacts of electricity raising price and the energy supply reducing shortage efficiency standards do not significant to influence the Taiwan residential electricity consumption behavior at sources.

It means that it won't influence Taiwan householders to use electricity or gas or any kinds of energy number to be reduced, even the Taiwan government energy suppliers sudden raise, any kinds of energy price and reduce to supply energy to satisfy Taiwan householders daily essential needs at homes.

In fact, Taiwan had improved gross domestic product (GDP) and it had raised employment rates recently. So, many Taiwanese has jobs to work, due to Taiwan economy had improved to be better. So, growth had also raised. The economy improvement causes many Taiwanese had enough jobs to work, due to new businesses are set up. Many consumers excit any kinds of businesses are invested to Taiwan from overseas or local investors. So, consumption is grown, the electricity consuming applicances are selected, as the household consumer focus grousp number if also influenced to be increased. So, Taiwan economy had improved to be better, it will encourage many electricity consuming applicances products are encouraged to excited to be selected to seel in Taiwan. Due to many different kinds of electricity consuming appliances are supplied to attract Taiwanese to choose to buy to bring to their homes for cooking, boiling water, or keeping rooms to be cooler or warmer temperature confortable feeling intention in winter or summer seasons. So, these electricity consuming appliances, e.g. rice cookers, heaters, air conditions, fans, bathing gas heaters etc. different home electricity consuming appliances will be increased to supply to satisfy Taiwan householders' needs. When they decide to buy any news electricity consuming applicances to bring to homes to use.

● Environment scientists' education message how to influence Greece householders home energy consumption behaviors from primary energy to change secondary energy

Finally , I shall indicate Greece, this western which will influence this country's householders have desires to do household energy conservation patterns or conservation energy consumption behaviors or energy conservation activities at homes. I shall explain the social economic variable, such as consumers' income and family size variation factor which can influence the different Greece family household members differences towards energy conservation preferences. IN addition, the variable, such as environmental information feedback and consciousness of energy problems are characteristics of the energy saver consumer.

Why and how can environmental pollution , environmental protection, energy conservation information message can influence Greece householders to choose to do energy use consumption conservation or less energy useful behaviors at homes. It is one interesting energy efficient use behaviors , due to Greece householders are influenced by energy conservation or environmental protection message.

In fact, scientists agree overconsumption of natural resources is a major threat to oue lives in earth. Environmental problems like greenhouse effect, ozone layer depletion, and acid rain effect are not any more problems of a specific region or environmental problem. Also, economic theory is indicated that in order to gain comfort and time households are becoming excessive energy users, neglecting the environmental impact of their choices.

Environment scientists bring these environment pollution message to influence Greeks (Greece householders) to change their energy consumption behaviors at homes. The environment scientists' message indicate that we are facing global warmth and natural resource and energy shortage challenges. Due to our Earth have limited natural resource numbers to supply to us to manufacture energy, but global population has been increasing every year. Thus, it is possible that we have energy shortage crisis. Also, manufactures are spending too much energy to waste to manufacture any products, the energy will cause air or water pollution in manufacturing process or drivers are driving their vehicles to pollute air on the roads.

Hence, environment scientists' message influence Greece householders began to consider these questions concern to reduce fossil fuel energy. Why do we need to Safety in using fuel and handle gas leaks? Why do we feel town gas smell? How is electricity located at electric station far away from town area? How to solve problems caused by the use of fossil fuels? How to reduce the use of fossil fuels?

Greece householders consider to solve the problems, the best way is to reduce thir used of fossil fuel. This helps prevent fossil fuels form being used up too quickly. Also, it helps them to reduce environmental problems because fewer pollutants are given out when less fossil fuels are used. Can human help to reduce the use of fossil fuels? Fossil fuels are mainly in power station. Although they use some fossil fuels for our gas cooker and car, it won't make much difference if I use less. Fossil fuel is not used renew primary energy. Most of energy Greece householders use come from fossil fuels, for example, the electricity we use is generated in power stations by burning fossil fuels. The buses they ride use diesel oil. Therefore, they can help reduce the use of fossil fuels by saving energy in Greece daily lives.

The actions that Greece householders can take such as: setting the air-conditioner to a higher temperature, walking instead of using lift, taking a short shower instead of a bath. This reduces the use of the hot water and thus the energy needed to heat the water. Thus, many people can help a lot to reduce our use of fossil fuels to avoid fossil fuel shortage risk occurrence.

Greeks (Greece householders) had been beginning to conern that they will face energy shortage challenge if they can not adopt more energy conservation actions. Because the Greece government began to bring negative environmental pollution and energy shortage challenge message if they often waste to use any kinds of energy, e.g. electricity , gas excessive number efficiency at homes. Then, they will be possible to fac energy shortage and environmental pollution challenge to their country in future one day. So, this energy shortage and environment pollution message has bring predictive negative worries to influence many Greece householder energy home users choose to reduce to avoid the waste of any kinds of energy use at homes.

So, their reducing energy use actions that had encouraged them to cause habits to avoid to waste excess energy to do any non essential electric appliances useful or consumption activities at homes often. Moreover, the environment protection and energy conservation message has changed many Greece householder to make decision and activities to change their lifestyle to b low living quality from high living quality. So, the environment protection and energy conservation message factor has much influential to change Greece household energy users' daily energy conservation or less energy use consumption activities at homes.

Greeks feel greenhouse energy can be environmental protection enegy. A greenhouse can trap heat in the sunlight and keeps the air inside the greenhouse warm enough for plants to grow. The glass roof and walls of a greenhouse let in sunlight but prevent heat from escape, this makes the greenhouse warm inside. Similarly, some gases in the Earth's atmosphere can trap heat from the sun and keep the Earth warm. This is called the greenhouse effect. The gases energy that can trap heat from the sun are called greenhouse gases. It is future one kind of potential primary energy to reduce environmental pollution new energy products for human consuming. So, environmental protection message influence them to consume greenhouse enegy at homes.

So, environment scientists' environment pollution message had influence Greece householders concern to apply seconday energy (environment protection) to replace electricity energy to use at home. They will change energy to use at home. The scientists' messages have more influential Greece householders energy change consumption behaviors at homes. The messages are as below:

There are different forms of energy, e.g. light, heat, sound, wind, water, electrical kinetic, chemical and potential energy. Some form energy is primary energy and it can not renew to use, e.g. light, sound, wind, water, fossil fuel etc. Some form energy is secondary energy and it can renew to use in possible, e.g. nuclear, electric charge battery etc. Why does human need to concern how to manufacture secondary energy? Because it is possible that our natural resource will be consumed all, thus we will face primary energy shortage risk. If human can invent any new form of man-made secondary energy to renew to use in order to avoid primary energy shortage to supply to use to use, then human won't only depend on our Earth natural resource energy supply numbers. We can invent any new secondary energy to renew to use again either replaces primary energy or instead of primary energy limit number supply.

What is energy change? For television energy change power case. Firstly, electrical energy changes to television power to be used by television itself, then it changes to light power, next it changes to light power. How to choose fuel form to use? Due to energy can change to different form of powers to supply different form of power advantages to supply to human to use, so it is possible that we can also invent any secondary man made renew used energy to change different form powers to supply us to use, e.g. nuclear energy changes to light or sound or heat form of powers ; electrical charge batteries changes to light or sound or heat form powers to satisfy our daily life needs.

The environment scientists' energy consumption education influence Greece householders concern how to change to use secondary energy to replace primary energy at homes as below:

For primary natural resource fuel energy example, different fuel has different feature, e.g. easy to burn, safe to use, gives out a lot of energy, inexpensive, produces little air pollution, easy to transport and store. How can we use in different channels, such as heating food, hot pat, driving vehicles.

For example, although coal is not expensive to cause electricity energy for past transportation tool, e.g. traditional coal energy train or our daily home cooking, but it has negative influence to environment air pollution. Hence, we ought to follow the primary natural resource energy's feature to decide how to apply what aspects of our life needs.

For example, if the country's people hope to reduce pollution when who use any kind of energy, e.g. US , Europe energy markets. The energy entrepreneur ought concentrate on manufacturing the kind of energy which can reduce environment pollution to be the least level to supply the country people to use, e.g. electric charge battery supplies to these countries' drivers to drive their vehicles on the roads, wind energy or water energy to manufacture electricity power supply to reduce air or water pollution ; or if the country people hope to buy the inexpensive energy to use, even the energy's quality and performance is worse, e.g. China, India, Hong Kong markets. The energy entrepreneur ought concentrate on manufacturing the lowest cost and enough supply of natural resource to manufacture the kind of energy to sell cheap price to these countries to use, e.g. China, Africa can accept to use e.g. gas, coal, fuel energy to use to compare developed countries people, e.g. UK, US; or if the countries people who hope to use energy which can easy to transport and store, e.g. light coal. The energy entrepreneur can choose to concentrate on manufacturing much coal to supply to the countries people to use, e.g. China, Arica Thus, to choose to manufacture which kinds of energy supply to the countries market people to use, the energy entrepreneur how decides to manufacture which kind of energy, it depends on which kinds of fuel advantages of the countries people most concerning.

What is energy meaning? It is defined a dynamic quality, it is a fundamental entity of nature that is transferred between parts of a system in the production of physical change within the system, and it is usually regarded as the capacity for doing work, and it is usable power (such as heat or electricity) or the resources for producing such power.

Why does secondary energy own investment worth? Because the different forms of primary natural resource energy will have supply shortage crisis, such as natural resources coal, gas, solar, wind, water, geothermal, biomass(organic material) etc. However, human can attempt to explore any undiscovered Earth or Space resource to manufacture any kinds of secondary energies, e.g. nuclear energy, electric recharge battery energy to supply to electric vehicle or space robots transportation tools to use or satisfy our daily life needs in future one day. So any kind of undiscovered secondary man-made renewed used energy resources have potential commercial worth to any energy entrepreneurs, it is possible that they can replace traditional primary energy to supply to human to use for our different aspects of life needs. In the future, the secondary energy demand will increase, when primary energy supply number has decreased form natural exploration. So, it will cause the effect of any demand of secondary energy product to be raised and prices to be increased in possible. Due to global population has been growing up, considerably China and India both countries populations have been increasing rapidly. Scientists predict there are more than 1.2 billion people worldwide will lack access to electricity, and more than 2.5 billion still use wood, charcoal to cook and heat in the future when primary energy has no enough number to supply to us to use. Hence, the fact that demand is this much greater than supply to make energy a prime market for further growth.

Although, secondary energy will have much investment worth, but energy like all other investments will carry risks. The internal and external risk factors include such as: policy is always changing to prohibit which do energy trading more easily between the energy exporting and importing countries, the secondary energy manufacturer itself own abilities to invent and to manufacture any kinds of secondary energy, improved technology can quickly make an technology obsolete, geopolitical rifts can happen overnight, the country's energy consumer (user)'s preferable choice to use which either kinds of secondary energy or secondary energy. So, it seems that (man-made) renewed used secondary energy industry can provide above-average returns, but it can also bring high risk commercial investment.

Traditionally, energy supply companies will apply those methods to operate energy providing businesses. For Shell,. Exxon examples, which had own gas stations, explore and drill for gas on their own. Other companies specialize in a part of the energy market, e.g. leasing oil rigs for example, or operating a pipeline. Energy supplying companies can choose to manufacture any kinds of energy to supply, e.g. trade oil, gas, coal, uranium, electricity etc. Any energy price and supply is demanded on the countries energy users' which kinds of energy most choice need or certain energy commodities to be chose to use popularly. For example, if US most people prefer to use secondary man-made renew used energy more than primary energy. Then, US energy manufacturers ought concentrate on manufacturing much different kinds of secondary man-made renew used energy to prepare to supply to its domestic US market in order to raise secondary energy price to sell in its country. So, the energy manufacturer's energy manufacturing choice, it is depend on which the country's people prefer to use which kinds of energy for their daily life needs.

However, scientists predict secondary energy market will have large market share, due to primary energy will have shortage to explore to supply in our earth and future energy consumers(users) prefer to choose to use more efficiency, less energy consumption, none environment pollution cause, cost effectiveness, renew to use of any kinds of energy. For example, the electricity recharge battery secondary man-made renew used energy is one kind of reducing air pollution power to push any electric battery vehicles to be driven to compare gas energy during drivers are driving their cars on the roads. They can reduce noise and air pollution and drivers can drive safely, who only need to buy one electric recharge battery to recharge in any electric recharge battery stations on streets when the electric recharge battery has no enough power to push their cars and they need to recharge their electric recharge battery drive when they had driven between one to two days. Due to primary energy, e.g. fuel , gas, the kinds of primary energies will have shortage to supply to global drivers to drive their traditional cars. Thus, the electric recharge battery or any undiscovered secondary energy will be future driving market needs. So, man-made renew used secondary energy, e.g. biofuel, hydro-electric, nuclear, will be one kind of efficient, clean, less pollution cause, cost-effective of energy to supply to our global vehicle market, even any other undiscovered new markets. Supposing they are popular to be used for electric vehicle market globally in future one day, then their prices will be decreased and constructed to average car requires up to 1,700 gallons of oil. Also supposing that making average computer requires more than ten times or weight to fossil fuels, every calories of food eaten in the US requires roughly then calories of fossil fuels. Hence, cheap energy will be one successful factor to influence future potential energy consumer (user) individual choice needs. Conversely, ion good economic times, people are more willing to travel, to buy products, and all of which success demand and low process for energy.

In the future, secondary energy will be the best choice to food production market. The modern food production system is essentially a success of changing fossil fuels into food. So, raising energy prices are almost higher food costs and even shortage for fossil fuels energy. If one day, one kind of discovered secondary man-made renew used energy can supply to any restaurants or homes to be used to cook at the cheap price, then the profit is very high for this kind of food production energy. Thus, future food production secondary energy consumption market is large and because the primary energy inputs for agriculture are higher than the energy outputs of the food. However, future secondary man-made renew used energy for food production system is only one part of whole energy consumer in food industry. The food production is related to whole food consumption market which includes: household cooking energy market, agriculture or vegetable, rice, fruit etc. foods farming machines energy market, food manufacturing factories market, food machine package market, transportation food delivery market, supermarket or fruit/food sale stores market. They must need any energy inputs to achieve the food production or food transportation or warehouse / stores electricity supply or cooking energy needs. Hence, these food suppliers relate to any whole food factory manufacturers, food retailers, food wholesalers, farmers and home/restaurant cookers, all of them must need to use energy to carry on their food producing or food cooking or food transportation activities every day in overall food industry. Thus, it seems that undiscovered any second energy demand will be increased, when the primary energy supply number is decreasing. Also, when people can accept to use secondary energy to replace primary energy to be used for any cooking, transporting food, manufacturing food, food retail stores or warehouse food delivery energy need activities. Then, the secondary energy price will be fall down to attract many food energy consumers.

Nowadays, the food industry energy may includes primary nature resource gas energy or electricity energy for house house families or restaurants cooking needs, food delivering lorry drivers driving needs usually. If future second man made renew used energy is invented successful popular to be used, e.g. hydrogen, electric recharged battery energy for electric vehicles or restaurant/home families cooking needs or food factories machine maufacturing energy needs. Then, the seconday energy will have possible to replace primary energy to be food industry energy market.

Wiley, composition services graphics indicated that global primary energy consumption had been increasing 30 billion tons from 1830 year to 510 billion tons in 2010 year as well as global population size had been increasing from 70 billion 1830 yeat to 510 billion in 2010 year. Thus, it seems that global primary energy consumption will be needed largely after 2010 year. If future global nature resource primary energy is explored full number and it had not enough energy number to supply global human to use. Then, it will being many people feel uncomfortable and inconvenient,e.g. Some countries won't have enough energy to supply transportion tools to be driven, some homes and restaurants won't have enough energy to supply to cook to eat or to provide restaurant clients to eat etc. daily activies, due to human's much activities which are needs energy supply. Thus, it seems that global primary energy comsumption will be needed largely after 2010 year.

Wiley, composition services graphics also explianed that why the primary energy consumption demand can be needed to achieve the same level to the global population size increasing in 2010 year. The graph showed these reasons why cause the same level of global population size and global primary energy consumpion demand which may include: The graph showed that after a nation is developed, its per-person energy use hegins to level off. In North Ameruca and Europe, where energy demand has remained flat, or fallen dightly, in each of the past few years. But the 1.3 billion people on those two continents are far outweighted by the 5 billion people in Asia and Africa, e.g. Chinese and Indian. who currently have more energy need to comapre average per man to North America and Europe per man, ensuring that overall energy demand will rise for years to come.

Wiley, composition services graphics also predicted that the growth in primary energy demand. China will have 4,500 million tons in 2035 year. India will have 3,000 million tons in 2035 year. Other developing Asia will have 2,000 million tons in 2035 year. Russia will have 1,500 million tons in 2035, Middle East will have 1,300 million tons in 2035, other rest of world will have 1,000 million tons in 2035. Hence, it implied that China will be the largest primary energy need country in the future.

China will be future the primary potential energy consumer market. The primary energy includes water, coal, wind, fossil oil, gas ,solar, geothermal energy, biomass (organiz material) etc. different natural resource primary energy. Otherwise, US, UK, Europe will be secondary energy potential need market. For example, electrical recharge battery energy will be raised demand to supply to any future new design electrical charge battery vehicles in US, Europe, UK markets.

Due to US, Europe, UK people concern environment protection, so they will invent many electric charge battery vehicles to consume electrical charge battery to replace polluted gas energy to avoid air pollution when the drivers are driving cars on themselve countries' roads. For example, second man-made renew used nuclear energy can be applied to rockets to pusch them to leave our earth to fly to other space far away and consuming nuclear energy will be cost efficient, and nuclear energy saving will be more when nuclear to spend long time to be used in any long time space journey. Hence, nuclear energy and electric charge battery secondary energy will be popular to be applied to vehicles and rockets energy needs in US, Europe, potential marketss, even our daily energy needs in global second energy market.

Who are your energy business's competitors (peers)? How do they compare? How have your energy business company performed cyclically? How to choose to manufacture to sell which kinds of primary or secondary energy product(s), either manufactures only primary energy product(s) or manufactures only secondary energy products or both? Which countries do you plan to sell your energy product?

Illustration by Wilsey, composition services graphiss showed that these natural resources to energy product the world's electricity percentage, such as below:

41% of coal, 5% of oil, 21% of gas, 13% of nuclear, 16% of Hydro, 3% other renewable secondary man-made energy.

Hence, coal will be future the major natural resource to produce electricity. The energy entrepreneur ought attempt to explore any coal resources, when who choose to supply electricity power to consumers for future energy consumption country markets.

Wiley, composition services also predicted that the expectation is that North America coal will supply the expectation is that North America coal will supply Asian demand, Us export terminals have a total capacity of 173 million tommes output. China will drive 16% of the nations total output. China will drive the sea-born demand for coal over for the forcessable future. Chinese energy consumption will grow more than 12 % between 1980 and 2009 years. Though, China heads global demand, India is growing faster in terms of coal imports. Much of the global coal demand will be supplied by Indonesia and Australia. Colombia, Russia, South Africa and Mongolia are also players in global export coal energy resources.

Hence, environment scientists' education messages influence Greece householders believe that secondary energy will be one kind of new energy product to replace traditional primary energy product for human energy consumption market global needs. Hence, it is right time any energy entrepreneur needs to research how to explore any undiscovered man-made renew used secondary energy products to avoid primary energy shortage crisis occurrence. Greece householders will be the highest population number to choose secondary energy to replace primary energy to use at homes. it means that environment scientists had changed Greece householders' energy consumption behaviors at homes.

In conclusion, different countries will have different factors influence how the country's householders energy consumption behavioral changes. Hence, it seems that any country's householders' energy use of consumption behaviors will be possible influenced by extermal environment factors influence. Also, every country's energy providers can attempt to find whether the country has what kinds of unique factors to influence its householders' energy consumption efficiency to increase or decrease in order to find the methods to solve the energy efficiency demand reducing challenges successfully.

Why environmental factor can influence
food shortage

Can artifical intelligence measure which the lowest environment pollution level influences food consumers' good wastage behavior rises ? What are objective indicators of standard of living and quality of life? Objective circumstances refer to the economic and material conditions which are important aspects of the standard of living and quality of life. In the assessment, eight different indicators were used: CPI, GDP per capita, shopping basket, household's expenditures, GFIC basket, poverty rate, income inequality and HDI. However, these indicators is one number measure. It can't measure anyone's psychological feeling, such as health, safe emotion. The challenge concerns whether environmental pollution factor, such as air pollution, water pollution can cause human's health to be poor, even goes down human's quality of life and economy loss. I shall indicate some evidences to give reasons to support my conclusion why I believe that environment pollution is a factor to cause human quality of life to be poor , even it can also cause economy will encounter loss too.

In general, measure of quality of life need include human's psychological feeling indicator. I shall indicate, Hong Kong, China countries air and water environmental pollution challenges how to influence these two countries' people quality of life to be poor, even, it will cause their economy loss. Nowadays, China and Hong Kong and India and Afria are encountering health problems arising from damage to lungs, heart and blood vessels. Hong Kong and India and Afria and China e.g. Shanghai city pollution is a significant cause of premature death from cardiopulmonary disorders. Present level of pollution cause injury to the immature developing lings of children and adolescents. This damage will lead to life-long health problems in many and a reduction in life-expectancy. Although, there is no evidence from analyses of trends in pollutants that pollution measures in recent years have reduced pollutant concentrations in a way which will benefits public health.

There are clear indicators that for some pollutants. The problem is worsening. In fact, air and water pollution is Hong Kong and China and Afria etc. developing countries' the biggest cause of social and environmental injustice. It harms not only citizens today, but because its transquenerational effects on the urborn and youngest members of the society, it will cause its will health effects well into the later years of this century, even environmental pollution challenge will cause these countries will encounter economy loss.

Human activities have created forms of air and water pollution, such as gases from fuels, uncontrolled emissons from fossil fuels and other chemical sources have long been recognized as a cause of ill health and premature death.

For example, in December, 1930 year, a dense fog affected the Meuse Valley in Belgium. Beginning on December, 3 date, the fog intensified over three days and was associated with laryngeal symptoms, chest pain, coughing, and breathlessness. Some patients showed signs of pulmonary oedema. Overall 60 deaths were attributed to the episode. After a long investigation, the cause was considered to be emissions from high sulphur fuels, including suplhur dioxide and sulphuric acid.

What is the current threat to health? the migigration of air polluton following the introduction of clear air has been followed by a period of unprecedented economic development creating new forms of pollution from the combustion of fossil fuels. For example, in constrast to the relatively large tar laden particulates from burning dirty coal which caused episodes like the London city, UK. Smog , traffic pollution now genertes fine with a different size and composition and gases,such as which may cause injury to the respiratory system and the effects of other pollutants. Such as particulates and drive the formation of the secondary pollutant ozone. The effects of pollution will therefore to some extent reflect genetic, environmental lifestyle and behavioral factors to develop these distance in a population together with the existing prevalence of diseases which may be polluted.

Hence, living in polluted urban environments is associated with increased levels of biological markers of inflammation compared with residence in a clean air environment. The damage is caused by air pollution manifests itself through a variety of common and recognized health problems, such as upper complaints heart and lung disease. Because of this, we can use statistical methods as well as clinical studies to detect the signal of changes in health problems and increased health care demands in the population. However, doctors had proved air or water pollution can cause these both curdiovscular or respiratory disease indirectly. Curdiovscular disease includes formation of arterial plaques, coronary artery, heart attacks, irregular heart rhythm, loss of heart rate variability, high blood pressure, stroke etc. disease. Respiratory disease includes inflammation of nasal, throat and tracheal airways with acute, lower respiratory tract inflammation and infection causing bronchitis, reduction long growth and function in young people. So, it seems environmental pollution can influence quality of life to human as well as environmental pollution and illness and poor health problem has close relationship.

On the other side, envionmental pollution can bring health risk, over it will influence social inequalities. Some researchers had found that the evidence has been compiled for six envionmental health challenges, such as air quality, housing and residential location, unintentional injuries in children, work related health risks, waste management and climate change. It seems human need to concern air and drinking water quality, waste management and climate change how to influence our environmental pollution challenge. Although, the evidence base on social inequalities and environmental risk is fragmented and data are often available for few countries only, it indicates that inequalities are a major challenge for environmental health policies. Irrespective of development status, environmental inequalities can be found in any country for which data are available. The valid for the exposure to environmental risk factor is also unequally distributed, and this unequal distribution is often related to social characteristics, such as income, social status, employment and education, even environment risk factor can influence human's quality of life.

(i) How environmental risk factor can influence different groups

However, human need to concern how environmental risk factor can influence inequally health outcomes to different groups. Such as, the first group is social determinants affect the environmental conditions of an individual and may contribute to the fact that specific individuals or population groups more often experience loss adequate or potentially harmful environmental conditions. The second group is the affected population groups could still be more exposed through e.g. the mechanism of education and health behavior. The third group is given socially disadvantaged groups could show more severe health effects of the social disadvantage is associated. The final group is social determinants affect health (what remains unclear is the relative importance of socially determined exposure to environmental risk factors). Thus, human need to concern our behavior can lead environmental pollution to influence poor health to alive. Even, we can not neglect how to protect our natural environment to be clean issue. Due to environmental unhealth poor issue can lead our bodies to be unhealth and to be ill and we have no health to work to influence our job inefficiency and low productivity if we often need to see doctor to raise workload to my staffs often. Then, our employers will be probable to dismiss and many unhealth employee will lose jobs and unemployment ratio will raise and DP will reduce, it will influence our economic growth . Hence, we can not neglect environmental justice and environmental inequity issue, e.g. indoor air pollution and occupational or exposure to environmental tobacco smoke pollution exposure to high traffic roads or to industrial plants pollution in our society.

(ii) How Afria country environmental pollution influences

Surprisingly, most of above countries , among of them, although Africa is a green and natural environmental country, but Africa has encounted poor natural environmental quality to influence it has poor quality of life to its citizen and poor economy growth to its society both. Why does Africa encounter this natural environmental pollution challenge? Afican have now two potential sources of pollution: consumption and production . This looks reasonable to Africa, since maintenance is completely dedicated to improving the environment, when production generates pollution only as a " by product". Capital implies the possibility of a country being trappical in an economent poverty trap by both a bad environment and low longevity. Some countries (or regions) may even experience other time, both environmental degradation and decay in expectancy. The fact that, in some cases, environmental degradation doesn't imply lower longevity may be due to the fact that economic growth might , at the same time, worsen environmental quality, but generate additional resources that can help increasing (or preserving) longevity. However, these is also evidence of countries where environmental degradation is associated with a reduction in life expectancy. It seems worsen environmental quality will influence any country's economic growth and poor quality of life both. For example, McMichael et al. (2004) identify 40 countries that experienced a loss in longevity between 1990 year and 2001 year (26 between 1980 year and 2001), they also support that the resulting world divergence in terms of life expectancy might be explained by ".... (the growing) health risks consequent on large-scale environmental changes is caused by human pressur, by both bad environment and low longevity, biodiversity and sustainable energy".

(iii) How human adult consumption and environmental quality influences future environment for human survival probability of life expectancy.

I shall assure human adult consumption and environmental quality has relationship to influence the future environment (green preferences) to provide human survival probability, it depends on inherited environmental quality. Thus, human will increase or decrease in the survival probability when we need a higher or lower life expectancy. In general, we depend on these environmental conditions to live, which include quality of water, air and soils etc. and resource availability, biodiversity, forestry, fisheries etc.

It is interesting to analyze different possible strategies to escape from the environmental poverty trap as well as factors that could push some economies back to a low equilibrium characterized. To research whether environment factor has relationship to influence human quality of life. We need to give idea of explaining whether environmental care has relationship to an uncertain lifetime. However, I suppose that an environmental kind of factor can be instead of being defined in terms of GDP per capita, capital accumulation etc. economic factors. Poverty is now related to environmental quality. It should be clear, however, I focus only on one specific mechanism lying behind environmental traps. Just as under development traps may be related to a wide variety of factors, ranging from financial to technological ones, including human capital accumulation and life expectancy. So, I should use this assumption to explain why it has relatively between environmental quality and life expectancy.

This " synthetic" indicator (YCELP, 2006) indicated environmental health is defined by child morality, indoor air polluton, drinking water, adequate sanitation and urban particulates and ecosystem vitality that includes factors like air quality, water and productive natural resources, A key ingredient of our setting is that survival until the last period is probabilistic and depends on the inherited quality of the environments. This survival probability affects the weight of the future environmental quality in human's utility function to achieve interest aim. Final stage, human will have optimal choices depend on life expectancy: in particular, a higher probability to be alive in the third period boosts investment in the environment and reduces consumption. In this case, a given country may be caught in a high morality/poo environment if low income is associated with a deteriorated environment.

John and Pecchenino (1994) were the first to introduce the possibility of multiple identifying, case for a poverty cause characteristic by poor economic performance and environmental degradation, however, life expectancy is assumed to be exogenous and plays no role in their model. Such as soils deterioration are the like, are all susceptible of increasing human morality (thus reducing longevity). So, the existence of both environmental performance and longevity, with countries being concentrated around two levels of environmental quality and life expectancy respectively. The two-way causes are between the environment and longevity. If the causal relationship between environmental quality and life expectancy involves the existence of an environmental poverty, characterized by both bad environmental conditions and short life expectancy.

Human life stage will encounter generations of three periods to get utility from consumption and environmental quality. During adulthood, when all relevant decisions are taken, adult can work and allocate their income between consumption and investment in environmental maintenance: consumption involves deterioration of the future quality of the environment (through pollution and/or resource depletion) when maintenance helps to improve it. The dynamics of environmental quality may also be affected by external factors on more resourced communities. The most importance, unhealthy physical environments across the region adversely affect everyone, ever though who are likely to be most concentrated in more burdened community which also have less social power to change those environments.

Why life expectancy and the environment has close relationship to influence quality of life? Life expectancy and environmental quality dynamics are jointly determined. Human may invest in environmental quality, depending on how much , we expect to live. However, environmental conditions affects life expectancy. In particular, some countries may encounter in a low life expectancy / low environmental quality. This outcome is consistent with stylized facts relating life expectancy and environmental performance measures. Some expects to live longer, who would be willing to invest more in environmental quality, because who feel which have causal link between life expectancy and environmental quality. However, environmental quality is a very important factor affecting health and morbidity: air and water pollution, depletion of natural resources and quality of life.

(iv) Why social and physical environmental factors have close relationship to influence economic growth or food shortage causing

I shall indicate reasons to explain why social and physical environmental factors have close relationship to influence economic growth, even human health of quality of life. The social and economic burdens of poor education, lack of affordable housing and less than self sufficient income affect, not just those individuals and families who have the fewest resources. The social gradient means that not only do whose in the bottom worse health outcomes to bottom of income group and the top income group whose will have poor quality of life influence. The higher rates of disease and disability and lesser productivity among many communities means a higher public and private burden of life years, particularly life expectancy once one reaches age 65. In recent decades, research and has increasingly shown how powerfully social and economic conditions determine population health and differences in health among subgroups, much more so than medical care. It seems that environmental factor can influence human's quality of life.

Los Angeles Country Department Of public Health (2016) indicated a country health rankings model, this department explained these three health factors can cause this health outcomes. These health factors include health behaviors (30%), it includes tobacco use, diet and exercise, alcohol use, unsafe sex; clinical care (20%), it includes access to care, quality of care; social and economic factors (40%), it includes education, employment, income, family and social support, community safety; physical environmental factor (10%) , includes natural environmental quality, built environmental quality. Then these factors can cause this health outcomes, such as morality (length of life):50% and morbidity (quality of life) :50%. SO, it implies that physical environmental factor can influence human's length of life.

So, on our social environmental problems result is from a complex interplay of a number of forces. An individual's health –related behaviors , particularly diet, exercise and smoking, surrounding physical environment and health care (both access and quality) all contribute significantly to how long and how well human love. However , none of these factors is as important to population health as are the social and economic environments in which human live, learn, work and play. We refer to these factors can be as the social determinants of health to influence our quality of life. How do social determinants affect our quality of life? In the late 19th and early 20th centuries, public health concentrated particularly on the physical environment. Improvements in, for example, clean water supplies, healthier housing, sanitation, workplace safety and safe food lead to sharp increases in average life expectancy . Also our quality of life needed to be concentrated on expanded access to medical care, resulting in further expansion. So, the poverty tap is now characterized by those elements, such as low levels of : (i) environmental quality, (ii) life expectancy and (iii) human capital.

In fact, environmental degradation can have a significant impact on human health. De Hollander et. al (1999) & Melse & De Hollander (2001) showed that estimates of the share of environment, related human health loss are as high 5% for high income countries, 8% for middle income countries and 13% for low income countries. Air pollution and exposure to hazardous chemicals are important causes of the related burden of disease in countries. The transport and energy sectors are major contributors to air pollution, when important sources of chemical pollution are agriculture industry and waste disposal. Opportunities for reducing environment-related health risks are considerable. The benefits of many environment policies in terms of reduced health care costs and increased productivity significant exceed the costs of implementing those policies. So, the impact of environmental risk factors on health are extremely varied and complex. For example, the effects of environmental degradation on human health can range from death caused by cancer, due to air pollution to psychological problems resulting from noise. So it implies environmental factor can influence our quality of life in our societies. However, many factors can also influence human's health of a population, including diet, sanitation, socio-economic status, literacy and lifestyle.

De Hollander et. al. (1999) & Melse and De Hollander (2001) showed that total burden of disease, with estimated environment-related share expenditure, mid-1990 year. The average income group has 15 daily/1000 capita, the middle income group has 20 daily/1000 capita, the high income group has 10 daily/1000 capita. As regards both total burden of disease and the health conditions related to environmental; degradation. The result indicates the environment –related share of the burden of disease is greatly dependent on income, with higher-environmental shares generally occurring in lower-income countries.

On the one hand, it seems the large environmental share of health problems is primarily, due to factors related to poverty, such as limited to access to proper food, housing, health care and drinking water. Environmental determinants of human health in developing or developed countries are related. On the other hand, those to the exposure to air pollutants (particularly in urban areas and chemicals in the environment than to poor living conditions. Also sources of human exposure to chemicals are many and varied. Chemicals can reach the environments, for example, through emissions from industries, anti-fouling paints on marine vessels, pesticides in agriculture, waste incineration and leakage from waste disposal sites. When emissions of chemicals from industries and other point sources of pollution have lead to poor quality of life, source of chemical exposure. Intensive agricultural production uses chemicals in pesticides and fertilizer and in feed additives and medication for livestock. Residues remain in fruit, grains, vegetables, meats and daily products, all of which can reach the consumer.

Other sources of chemicals in food include bio-accumulative chemicals in the environment, such as heavy metals and persistent organic pollutants, which can be found in fish, meat and dairy products. So, environment pollution can influence human need to eat bad or unhealthy food to cause we have poor quality of life to live, such as the high income group or middle income group or low income group of families in our societies fairly. Other human health risks that have recently received considerable attention include unsafe livestock feeding practices through which toxins reach the food chain unintentionally. Dioxins that have accidentally contaminated poultry feeds that contain diseased animal remains can cause the so-called " mad cow disease" in livestock which has been linked to a new form of disease. The effects on health from exposure to chemicals and air pollutants vary from allergies to cancer. Although, the link between exposure and disease is often not clear, Even at low exposure levels, urban are pollutants can cause, asthma, allergies, respiratory diseases and cardiovascular disease if the exposure is continuous or long term. Heavy metals have been shown to cause neurological disorders and various cancers. In addition to , physical diseases, environmental contamination can also cause psychological problems. Noise, one of the determinants of the quality of urban life can have an impact on human health, decreasing the quality of life and potentially contributing to depression.

For Ireland, UK country example, this country politicians and policy makers believe the role of environment can be used to measure quality of life, concerning on either in its own right or relative to economic and social aspects of quality of life. Agreement on what measures quality of life and how it can be measured by the role of environment, not just in Ireland, but everywhere. The conventional approach is used for policy has been to use measure of gross domestic product(GDP) or regional valued added. However, it is acknowledged that such conventional economic measures have only a partial relationship with societal wellbeing. To the extent that economic measures are related to public products and consumption, there are also pressing issues in relation to public products and the sustainability of economic growth. However, the role of environmental factor can influence resource use and human's behavioral consumption.

Aspects to quality of life other than income include the environment, freedom, health, working condition, leisure, social and family relationship. Economists don't deny that these factors do play a role in quality of life. However, environment factor can be one role to influence other factors to influence our quality of life to be good or bad effect.

For example, locations which might be desirable as paces to life (in terms of income earning opportunities or other factors) were also likely to have higher costs of living, particularly with regard to house prices or health or unhealthy air/water pollution of environment situation of the place to provide human to live. Alternatively, social indicators are based on normative ideals of literacy, low rates of premature mortality or a quality environment. Other measurement of people's personal evaluation of their quality of life, much depends on personal expectations and experience.

(v) How environmental factor can influence any country's house price and food shortage causing

I shall indicate that why environmental factor will influence any country's house price. For Ireland example, citizen average incomes and higher in the east of the country, house prices are lower in the west, who are also more able to afford a property of choices. There are more opportunities to purchase houses, where people own their own houses, who are more likely to have benefits from an appreciation is its value and to consequently perceive a higher degree of health. Generally, levels of property appreciation have been higher in the east. Unfortunately, young people and the more economically active segment of the population are more likely to be faced with rising entry level house prices and the prospect of large borrowings. So, the quality of life, such as education, crime and access to healthcare and living environment are not uniformly better in the west or the east regions. Indeed, many measures of social disadvantage are at their worst in the west regions. Some indicators of environmental quality are , indeed better in the west regions, but there are others, such as drinking-water quality or recreational access that are often worse.

Comparisons can often be reduced to an urban-rural dimension rather than a regional one. Factors such as incomes, house prices, crime levels, air pollution and congestion are all likely to be higher in urban areas in Ireland city, UK country. Why environment and housing price has relationship in Ireland to influence quality of life to its citizen? If Ireland's regional development policy is successful , it will bring with it greater competition in the housing market and greater pressures on the environment in Ireland. Because the forest will be decreased to build house, the natural environment will become wood and steel and stone of housing built environment. In fact, it appears that there is a fair of amount of agreement on the relative rating of factors influencing quality of life. Ability to own one's home and security of income were needed, but respondents also placed almost equal important on clean air and drinking water, low crime were differences. The Ireland's rural respondents appeared to place a slightly greater emphasis on key natural environmental attributes, when urban residents valued absolute incomes and social or leisure activity rather more.

In this respect, the analysis identifies three components to Ireland people of quality of life, each of which was evident in all three locations. There components can be broadly described as domestic security, social/leisure and aspects of the planned environment. The first of these includes indicators, such as security of income, absolute income, house ownership and low crime. As this component includes air and drinking-water quality, it suggests that these indicators may be associated with personal health and well-being. When the planned environment component includes those attributes that affect quality of life over which the authorities have a direct influence, for instance, a clean environment, traffic and reducing vehicle numbers on the roads in busy time.

(vi) How environmental pollution can influence social welfare and cause food shortage

Environmental quality has an undefined impact on quality of life and various indicators are used to show regional variations in aspects, such as water quality . There are many measures of environmental quality , but is only for quality of life. Moreover, the measurement of societal welfare is important. Societal welfare is not simply , the sum of the parts, but varies depending on the individual in which people find themselves at any time in their life. In principle, it should be possible to apply weights to each element of societal welfare, but as preferences for each of these vary within the population. In the absence of a method with which everybody is satisfied, GNP and GDP are typically the most popular used measures for quality of life or standard of life. But, these are problems with the data itself to measure quality of life because quality of life is feeling or satisfaction of level to the country's citizen and it can not be seen by numbers or statistic method. For example, GDP ignores household production, such as the effort that goes into the rearing of children, the benefits that this provides for society and the public expenditure that is avoided. Neither are costs treated equally with the benefits. GDP counts all economical activities irrespective on pollution appears to increase. GDP even through it is a degree of double counting. Otherwise, environmental products are good to be measured to quality of life. For example, many environmental products are unpriced. Consequently, environmental products that people value, or which are critical to the sustainability of development, are abused or depleted because of their public products have good characteristics and the absence of a market price signal.

Environmental economists try to work within the economic model to measure quality of life. Rather than questioning the link between utility and consumption or choice, the preferred approach is to add an element into the utility function that represents the value of environmental products or the stock of natural capital. By one means or another , the preservation value of these environmental products is estimated in terms of willingness to pay to protect the environment or as willingness to forego other products in return. It seems the quality of people's environment can be represented by objective indicators. At another, their interpretation will vary and can be represented by subjective indicators.

Objective indicators come in two forms: (i) economic indicators and (ii) social indicators. The former depends on an ability to select the products and services that are desires, in other words, the satisfaction of preferences . The economic argument is that people select the best quality of life, who can obtain commensurate with their resources and personal desires. By comparison, social indicators are based on normative ideals on what could be considered the food life. For example, would be infant morality, literacy, crime rates and social indicators are objective measures.

Both have guided, much of the research on quality of life, particularly concerning with the urban environment. Quality of life can include natural a significant influence on local quality of life, for instance, natural beauty spots used for recreation.

How artificial intelligence impacts energy
consumers using behaviours

Nowadays, many countries began to educate citizens who have responsibilities to use energy at homes or offices or public places or any indoor environments in avoiding to do energy wastage behaviours or misuse energy wastge attitudes as well as teaching them have responsibilities to protect their earth's natural environment to reduce air, water pollution in order to void rising temperature to bring globl warm challenge to influence our quality of life to be poor, even facing death threat, due to our natural environment is damaged and polluted by our energy wastage behaviours.

In fact, I feel the energy wastage eduction is not one efficient or effective method to persuade every energy consumers, such as householders, office workers, factories workers, any entertainment places workers or enjoyers, such as cinema service staffs, shopping center staffs etc. entertainment places to reduce to use any electricity for light or any entertainment aims to consider themselves working environment or entertainment environment to satisfy, e.g. cinema movie to satisfy customers' needs. For example, private vehicle drivers, public transportation tool drivers, householder energy users, businessmen energy users who still only consider themselves passengers comfortable aims, e.g. spending much electricity often to turn on light in buses, taxi, cars, trams, trains, underground trains in morning or afternoon time. So, these public transportation tools are popular to waste electricity because they expect their passengers to feel comfortable in summer , so they will often turn on air conditioners to keep colder in summer or turn on warmers to keep warmer in winnter all the transportation working hours. So, these drivers are doing energy wastage behaviours. Moreover, these private car drivers only expect to feel comfortable , so they will open air conditioners to keep colder in summer when they are driving cars, even they are stopping cars on the road. So, they are also waste energy.

Hence, they will be negligent to consider how to use energy in efficient attitudes or energy saving behaviors in order to avoid energy shortage challenge occurrence. However, since (AI) technology began be popular to be accepted to use by human. (AI) scientists began to carry on researching how to apply (AI) technologies, e.g. big data gathering , robotics to assist human to adapt or learn to use any kinds of energy in efficient and no wastage attitudes or using behaviours habitually.

I shall explain how to apply (AI) technology to assist human to adapt to use energy in order to avoid to do energy wastage behaviours easily to every energy users as below:

In consumer psychological view point, the behaviours of individuals can have a standard rational choice model, in which people, such as energy users objectively weigh up the costs and benefits of investing time and money into " greening" their homes or offices or any working places or entertainment places or transportation tools being more energy efficiently. So, the social, cognitive and behavioural factors are important in explaining why many energy users, such as householders, vehicles owners or public transportation tool drivers, office workers, businessmen who are neglient to avoid to spend much excessive energy to drive their vehicles on the roads , to turn on lights in offices or any working environment or entertainment places or at homes all days. When they feel that they need more enjoyment, raising productivity, raising service performance to satisfy customers' needs. So, when they weigh economic benefits and cost. They will choose to use more energy to achieve their profit growth or customer number growth or improving quality of life intentions.

Hence, it also explain why education method is not effective to achieve energy saving aim for every energy users, e.g. it has no reward to compensate to their losses, when they choose to reduce energy consumption to cause that they have economic losses. So, it seems any country's government or schools energy saving education method which won't achieve the best energy saving consequency nowadays.

Why does (AI) influence energy users to reduce and use more energy in order to achieve their energy -saving habitual impact easily than education method? I shall explain as below:

(AI) technology can be one auto-manual tool to help householders to protect their homes to be more green environment and be more energy efficient. For example, householders can install (AI) auto-energy efficient measurement tool to record whether they will spend how much money for energy . e.g. electricity , gas consumption at home every day. So, they can know whether they will pay how much money for electricity or gas fee. (AI) auto-energy efficient measurement tool can also change householders' energy consumption behaviours to save more electricity or gas when they discover that the day' electricity or gas using number is excessive to cause they will be pay more extra electricity or gas expenditure on the day. Then, they will find whether why or how or what reasons cause them to spend excessive electricity or gas energy at homes, then they will change their energy wastage behaviours to save energy more easily. So, their energy -saving behaviours are influenced by the (AI) auto-energy efficient measurement tool's daily electricity or gas using record at homes.

So, (AI) auto energy efficient measurement tool can help householders to save energy and money when they need to use electricity or gas energy at homes. But making the kind of improvements that have these effects is not always simple, they usually require some planning, time to prepare to adapt how to do avoiding energy wastage behaviours at home habitually . So, (AI) auto-energy efficient measurement tools can focuse on what householders might be able to do and further encourage the uptake of energy efficiently measures as well as it might be able to motivates householders to act through restructuring existing incentives and using collective rewards.

When, they discover that the (AI) auto energy efficient measurement tool shows either electricity energy or gas energy or both using energy number is excessive too much to compare the normal energy using number on the day suddenly. Then , they will attempt to find what reasons influence their energy spending number is excessive on the day, in order to change their energy using behaviours or habits and they will be more acceptable to adapt to do energy saving behaviours because they can earn energy expenditure saving rewrd and money saving reward in order to avoid further the excessive energy using number to be increased to pay more electricity or gas energy expenditure , due to they often do unnecessary energy using habitual behaviours at homes.

Hence, (AI) auto-efficient energy measurement tool will have much effort to persuade householders to choose to do energy saving behaviours habitually at homes, due to it can provide the more acceptable number concerns their daily electricity and gas energy using record at homes to let them to know how and why their energy expenditure changes to spend more suddenly in order to let they can understand the reasons why and how cause their needs to pay extra excessive energy expenditure at homes.

The main important reward is that the householders can be encouraged to measure their energy using number and find the reasons why and how their energy using behaviours cause their extra excessive energy using number on the day. Then, they can find what the factors are to cause their electricity or gas energy using number to be increased suddenly on the day and change their energy using behviours to avoid the energy using number to be continus increased in order to avoid to pay extra excessive electricity or gas fees on the month immedicately.

6.1 How to apply (AI) technology to improve energy efficiency and better climate change and the security of energy supply as well a resource efficiency?

Increasing energy efficiency involves using a reduced quantity of energy to achieve the same or improved product, process or sevice. It is generally measured in a physical unit as the ratio between energy output and energy input. Similarly, resource efficiency refers to the ability to use a reduced quantity or volume of resources to produce the same or an improved service or product and it is measured as the ratio betweenn useful material output and material input, both measured in physical terms (Dahlstrom and Ekins, 2005).

Hence, if it was only (AI) technology can increase energy efficiency or reduce resource effifiency to improve service performance. Then, it will reduce energy wastage. So, it bring this question: How to apply (AI) tool to reduce resource consumptin indentified by analysis of historical resource efficiency?

It presents an historial analysis that seems have relationship between energy and resource efficiency improvements and resource consumption across a number of different sectors of activity, including iron, and steel production, electricity generation from coal, oil and natural gas and motor vehicle travel.

So, future(AI) technology needs to fight social and behavioural barriers to energy efficiency in the housing sector. If future (AI) technology can improve energy efficiency for home renovations and it can consider the social factors. It is a qualitative investigation technology of the decision making process guiding to teach householders hoe to use overall energy was reduced by the householders' house renovation. It will also bring another question: How can (AI) technology can help householders to do decision making to reduce overall energy consumption by householders' houses renovation, such as reducing energy using when the householder needs to renovate whose house's design, e.g. extensions and additional bedrooms or bathrooms . So, future (AI) technology can be needed to help low income householders to increae energy efficiency and reduce energy consumption when their homes need to renovate whose houses' design ,e.g. extensions and additinal bedrooms or bathrooms or bookrooms or children toy rooms at hoomes.

Due to low income householders were concerned about energy consumption for environment and economic reasons, upfront costs rather than life-cycle costs were considered more important when the low income householders need to renovate to extend extra bedrooms, bathrooms, studyrooms to buy extra electronic appliances to install them in these rooms to use. Then, they will be concerned energy awareness how will be more consume when the low income householders choose to renovate their homes design to extend more rooms to feel more comfortable, or large size, but they also need to consume or use more electrciity or gas energy for extra electronic appliances in these rooms possible as the same time.

Hence, future (AI) technology needs to assist these low income householders how to reduce or avoid to ue extra more energy, when they renovate their homes' designs to cause to need to buy extra more electric appliances to use more electricity or gas energy at homes. Moreover, future (AI) technology ought have effort to help any countries' buildings to be efficient energy saving buildings to be efficient energy saving buildings to help householders to use less energy to live in their builsing efficiently. When the country's overall buildings can use energy efficiently , it won't only bring energy saving benefits, even it can bring the country's economic cost to be reduced , due to any building' energy efficient using high technological method. So, future (AI) energy saving technology will concentrate on how to help any buildings to use energy efficiently , in order to achieve energy -saving efficient buildings to let the householders and office energy users to either live or work in energy -saving efficient buildings to avoid energy wastage aim.

Hence, future (AI) energy -saving technology needs to focus on how changing energy users' energy wastage behaviours to energy -saving behaviours. How to apply (AI) energy-saving technology to assist energy users change their behaviours to spend unnecessary excessive energy to use habitually daily.

I recommend that future (AI) robotic cans be such as energy-saving machines to help any factory workers to cooperate to work to achieve efficient energy -saving aim, but they can also raise productivities. So, factory robotic are as learning tools, allowing factory energy users (factory workers) to teach themselves how to use less energy to achieve the productivity won't be decreased intention. So, when the robotics and factory workers work in the factory environment together. The robotics can give feedback to let these factory workers how to cooperate to use lesser energy to work efficiently in factories.

Other information and advice achieving better understanding and control of energy use in factory. So, future factory robotic machine men are such as teachers teach students in classrooms or trainers provide training to train trainee in factories. It means that robotics and factory workers can learn how to understand to do every working steps to avoid to spend extra excessive energy , but they can also raise productivities as the same time in factories.

So , future (AI) robotics will be demanded to invent to be one energy-saving machines to assist factory workers to use lesser energy to manufacture any products in manufacturing process, but they can also have productivity and efficiencies won't be reduced in the efficient team work method. So, every factory robotic machine mman is needed to be designed to own the advanced manufacturing technological skills or manufacturing methods to assist the factory workers to manufacture the kind of products in team work together in order to shorten time and using the most efficient manufacturing methods to achieve and produce the best quality products and the highest productivity in energy-saving working environment in factories. For example, when every team watch factory's factory workers who need to operte with ten workers per team in the watch manufacturing factory. One robotic machine with ten workers per team will need to raise to manufacture at least fifty watchs number per hour to compare only ten workers per team can manufacture the maximum fifty watchs number per house, when the robotic machine participates to every team to work together.

The robotic machine must need to help them to use lesser time and electricity energy to manufacture more than fifty watches number pe hour in order to achieve long tem energy saving and time saving and efficient raising productive economic benefits to the watch manufacturing company. Hence, the watch factory's every watching manufacturing robotic machines can encourage the watch manufacturing firm to choose to use them to assist every team watch manufacturing workers to work in order to achieve high efficient productivities, high quality of watch manufacturing, reducing every watch manufacturing , reducing every watch manufacturing time and the important intention is energy -sving efficient benefit to reduce to spend more extra excessive electricity for long term expenditure.

So, in the future , every robotic machine will need have these benefits to satisfy manufacturers' every -saving needs in their participative manufacturing process in order to achieve energy expenditure to reduce for long term economic benefits to persuade them to use these energy-saving efficient robotics in factory attractively.

The relationship between consumer psychology and behavior

Can every consumer himself/herself personal psychology influence whose consumption behavior how to do final purchase decision? Why and how can psychological factor influence consumer behavioral change? Must psychological factor be the major influence to cause any consumers to make final purchase decision? Can other factors influence psychological factor to influence consumers final purchase decision?

I shall indicate online shopping method, how online shopping consumers are influences to make final purchase choice from their psychological changing influence. In fact, it may have different factors to influence consumer behavior towards online shopping. Usually online buyers will choose to buy any things, when they stay at home in preference. They feel turn on home computer to shopping in preference, due to They feel turn on home computer to apply internet channel to buy any things will be more safe to compare to use public libraries, shopping centers, schools, office computers' internet to buy any things. It is one good example to explain why online shoppers usually like to buy any things to turn on home computer internet to spend much time to choose to buy any things. Safety feeling psychological factor is one important factor to influence online shoppers prefer to turn on home computer to apply internet channel to buy any things . It is possible that they feel public places' computers are Insafe and their visa cards personal data will be theft by any one easily.

Due to online shopping method has risk to bring any online shoppers' visa card personal data loss or theft. So, safety psychological factor will influence many online shoppers choose to apply internet transaction tool to buy any things at home. Unless, the online shopper has no computer at home or whose computer has no internet installation to his/her computer. Then, he/she only chooses to use public computer internet to buy any things. Hence, it explains that safety psychological factor can dominate or control any online shoppers' purchase behaviors to choose where is the most suitable place to use computer internet to carry on online shopping buying action. Also, it explains safety or privacy factor (avoiding visa card data theft) which will be one important factor to influence online consumers' preference purchase place choice (turning on the place's computer and applying its internet to carry on online shopping actions). Even, time factor and convenient both factors will not influence general online shoppers' using home computer and internet tool purchase behavior. For example, one student is staying at whose school library place at this moment, his school library's computer has installed internet too, he plans to buy one book for his reference at this moment. Although, he can apply this school library's computer internet to buy this book immediately. But, when he feels that he can buy this book immediately , if he decides to apply his school's library computer website to find any book shops , whether they have this book to sell to him. Possibly, he only needs to spend more ten minutes, or less than ten minutes to buy this book. SO, he does not need to spend longer time to catch any public transportation tools to go home and he also needs much time to turn on home computer to search any online book shops' websites to buy this book from himself home computer internet channel. However, it is due to privacy and safety psychological factor influence where is the most suitable place choice to buy this book from internet channel. SO, it explains that why psychological factor will be more important preference to compare time and convenient factors to influence online shoppers' (online shopping place) choice.

How psychological factor influences women /men (females/male) shopping place) choice. Shopper behavior has focuses on individuals and the factors that impact their decisions to spend their resources on consumption –related items. Female and male's consumption model or behavior has different, due to their sex is different. So, their psychology are also different to influence their shopping behaviors in possible. For example, women think differently from men because there are biological , neurological and behavioral variations between the brains of men and women . This differences in turn make an impact on their shopping behavior. When, men will load themselves with sufficient information of a product or service through internet, advertising, reviews. Otherwise, women would try to get benefit from others' shopping experience by asking peoples' ideas before they choose to buy any products.

Retail is the dominion of women and shopping is an action frequently seen as complementary to female role. IN general, female like to go to supermarkets, or any food shops to buy any kinds of food more than male. Women go on shopping to purchase both essential and discretionary products (daily living useful products), more than male. Females (women) are considered to be the most potential household consumers, as it has almost 80 percent of the domestic expenditure is spent by female to every householder (family member shoppers).

Any family member mother or sister will be the influential role. Female is the influential member , she can make final purchase decision in any householder (family) usually. It is possible due to any family member father or son has more confidence to believe mother or sister whose final purchase decisions are more accurate judgement before they pay to buy any things for home to use or eat. Hence, this female's accurate shopping judgement influence role or feeling which is one major psychological factor to explain that why householder female member can be the final shopping decision maker more than male member in any householder, because they are felt to be one confident home shopper to every male family member. In general, they are felt to make more reasonable or accurate judgement to buy any kinds of products or food for home to use or eat to compare male family members' shopping judgement. Nowadays, it also explains why " confidence" is one important psychological factor to influence female, which is the major influential final shopping decision making more than male at any homes.

How can psychological factor influence either online or offline (visiting shops) shopping behavioral choice? Traditional shopping (visiting shops) had been one kind habitual consumption model or method, but internet shopping method had been also popular to be accepted to young people, e.g. student shoppers, even old people who also accepted to use internet to shopping. However, psychological factor is one important factor to influence their shopping method changing.

Online shopping is real fact in recent years. It will bring innovative and high technological shopping method to change traditional visiting shops shopping method as soon future. IN nowadays shopping environment, consumers like to buy any things conveniently in short time. Busy working and living factor influence their shopping feeling to be changed. IN general, consumers do not like to spend much time to visit any shops to choose any products and make final purchase decision. They like to make purchase decision in short time immediately. Due to internet is convenient, consumers they do no need to find where the shop can sell the product. They only need to turn on computer and enter any websites to search whether which online shop provides the kind of product to let them to bur from internet. Convenience and fun and high technological internet shopping method will be one kind of attractive shopping method to replace traditional visiting shops shopping method to change any one of consumers their traditional and habitual shopping model (method) or consumption behaviors daily.

The reason may be busy working time psychological factor to working people as well as busy learning time psychological factor to students. Convenient shopping factor, high technological invention of electronic shopping method factor. Many students and working people hope to spend less time to choose whether any product is the most suitable product to buy. They do not want to spend much time to search which shop can sell the product. Working and learning time is more important to compare shopping time. Hence time psychological factor will influence many young consumers choose to buy any things form internet channel. Otherwise, old people are retirement, they have much time to relax or entertain. SO, usually old people do not like to apply internet to buy any products at home. Because they have much time to visit any shops to go to shopping . Hence, psychological time factor will influence the shopping choice behavior difference between young age consumers and old age consumers.

Why can product knowledge psychological factor influence consumers product choice behavior? If the consumer feels the brand's any products have better to quality to compare other brands' products. He has confidence and product knowledge to use the brand's product. He has good past purchase experience to use the brand's products, because his useful feeling is good and he feels the brand's product can satisfy his need to use it. His past useful experience of this brand's product psychological factor will cause his future preference purchase choice , when he feels need to buy same or similar product in marketplace. So, it implies that when the consumer has good purchase and useful experience to the product, he will repeat to buy the brand's product again. Otherwise, when the consumer has bad purchase and useful experience to use the product before, he will lose confidence to buy the brand's product again. Hence, it explains why whether the consumer's good or bad purchase and useful experience psychological factor will influence whose purchase choice either repeat purchase or not repeat purchase again to the brand product.

Marketing message (unique value) and advertising campaigns influence psychological factor, it can influence consumers' confidence to use the brand's any products, even increase sales, because if the brand product seller can build good image to let consumers (public) to know its products existence by television, radio, newspapers, internet advertising channel. So, it seems that public advertisement promotion method can build good image to any product, it can achieve good image psychological factor to persuade any customers to attempt to buy the product to use or buy the food to eat. It is one building good image to let consumers to know the brand existence in the marketplace, it can bring message concerns that whether what kinds of products it is selling, how much price, discount it charges, what advantages of its products. All of these advertising message may persuade any customers to make first time purchase choice or repeat purchase decision, when they are watching its product advertisement from television, listening its advertisement from radio, reading its advertisement from newspapers or magazines etc. different advertisement channels in any time easily.

Any one of its advertisement message can let any audiences to remember the product's shape, color, size in their brains in long time. SO, long time advertisement memory to the product , it will bring positive consumption image influence to any audiences as well as persuading them to make final purchase choice to the product more easily.

Hence, advertising can shorten any consumers' purchase process or choice time to the product. Consumer behavior refers to the selection, purchase and consumption of products and services for the satisfaction of their wants. There are different processes involved in the consumer behavior. Initially the consumer tries to find what commodities, he would like to find what commodities, he would like to consume, then he selects only those commodities that promise greater utility. After selecting the commodities, the consumer makes an estimate of the available money which he spend. Lastly, the consumer analyzes the prevailing prices of commodities and takes the decision about the commodities he should consume usually. Hence, one attractive advertisement can influence many consumers feel that they do not need to spend much time to compare different kinds of similar products (product selection time), because the advertisement had attract their consideration to make final purchase decision. Hence, one attractive advertisement can shorten any consumers' (product selection time) to compare different kinds of products in order to make final purchase decision. Because it has build confidence to let many audiences feel the brand's products are worth to buy to compare other brand's products.

On conclusion, above these consumer behavioral purchase cases explain why consumer individual psychological factor can influence his/her final purchase decision. It implies that it has close relationship between consumer psychology and behavior. How the consumer's feeling to the product, it will cause his purchase behavior either selecting purchase it or selecting another product to replace it effect. SO, psychological factor is one absolute important factor to influence consumer behavior.

www.ingramcontent.com/pod-product-compliance
Lightning Source LLC
LaVergne TN
LVHW060038210926
848481LV00013B/3891
* 9 7 9 8 8 8 7 4 9 4 6 3 0 *